RUSSIAN
TALES AND LEGENDS

Russian
Tales and Legends

Retold by
CHARLES DOWNING

Illustrated by
JOAN KIDDELL-MONROE

OXFORD UNIVERSITY PRESS
OXFORD   NEW YORK   TORONTO

Oxford University Press, Walton Street, Oxford OX2 6DP

Oxford New York Toronto
Delhi Bombay Calcutta Madras Karachi
Petaling Jaya Singapore Hong Kong Tokyo
Nairobi Dar es Salaam Cape Town
Melbourne Auckland

and associated companies in
Berlin Ibadan

First published 1956
Reprinted 1960, 1964, 1968, 1973, 1978

First published in paperback 1989

British Library Cataloguing Publication Data
Downing, Charles
  Russian tales and legends
  I Title II.Kiddell – Monroe
  823.'914 [J]
ISBN 0-19-274144-6

Printed and bound in Great Britain by
Richard Clay Ltd, Bungay, Suffolk

Contents

HEROIC POEMS
(Byliny)

FOLK-TALES
(Skazki)

HEROIC POEMS

(Byliny)

Volga

As the red sun sank behind the dark forests and the broad sea, and the stars in their myriads spread over the clear heavens, then was Lord Volga Buslavlevich born in Holy Russia. And when Volga had grown to the age of five years he walked over the land and Damp Mother Earth trembled beneath his feet, beasts fled to the forests, birds started up among the clouds, and the fish in the blue seas scattered in terror. And Volga Buslavlevich went and learned wisdom and skill and all the divers tongues of men for the space of seven years, and when he had attained the age of twelve years, he assembled a trusty druzhina in Kiev —thirty bogatyrs but one, and the thirtieth was he.

'Good and brave druzhina,' said Volga, 'listen to your elder brother, your ataman, and do as I command. Weave me nets of silken cord, set them in the dark forest on the damp earth, and hunting for three days and three nights, catch for me martens, sables, foxes, white hares, tiny stoats, and all manner of wild beast.'

The druzhina obeyed their elder brother, their ataman, and did as he commanded them. They wove nets of silken cords, and setting them in the dark forest close to the damp earth, they hunted for three days and three nights, but not one single beast fell into their snares. Then Lord Volga

3

Buslavlevich turned himself into a great lion, and bounding over the damp earth into the dark forest, he headed off the martens, black sables, foxes, white long-legged hares, tiny stoats, and all manner of wild beast, and drove them all into the silken nets his men had woven.

And when they were in the famous town of Kiev, Lord Volga Buslavlevich summoned his trusty druzhina and addressed them thus:

'Good and brave druzhina, listen to your elder brother, your ataman, and do as I command. Weave me nets of silken cords, set them in the dark forest in the tops of the trees, and hunting for three days and three nights, catch for me geese and swans, hawks, white falcons, and all the small wild birds.'

The druzhina obeyed their elder brother, their ataman, and did as he commanded them. They wove nets of silken cords, and setting them in the dark forests in the tops of the trees, they hunted for three days and three nights, but not one small bird flew into their snares. Then Lord Volga Buslavlevich turned himself into the great Naui-bird, and flying into the sky up to the clouds, he headed off the geese and swans, the hawks and white falcons, and all the small wild birds and drove them all into the silken nets his men had woven.

And when they were again in the famous town of Kiev, Lord Volga Buslavlevich summoned his trusty druzhina and addressed them thus:

'Good and brave druzhina, listen to your elder brother, your ataman, and do as I command. Weave me nets of silken cords, take your sharp axes and make me a ship of stout oak, and going down to the blue sea for three days and three nights, fish for the salmon, the pike, the roach, and the precious sturgeon.'

The druzhina obeyed their elder brother, their ataman, and did as he commanded them. They took their sharp axes and made a ship of stout oak, and going down to the blue sea they fished for three days and three nights, but not one fish swam into their nets. Then Lord Volga Buslavlevich turned himself into a great pike, and flashing through the waters of the blue sea, he headed off the salmon, the pike, the roach, and the precious sturgeon and drove them into the nets his men had made.

And when they were again in the famous town of Kiev, Lord Volga Buslavlevich summoned his trusty druzhina and addressed them thus:

'Good and brave druzhina, listen to your elder brother, your ataman. Whom shall we send to the land of the Turks to know the Sultan's mind, and what his intentions are, and whether he thinks to invade Holy Russia with his men? If we send an old man, he will go slowly and we shall wait long for news; if we send a man of middle age, the Turks will make him drunk on wine, and we shall still wait long for news; and if we send a young man, he will dally with the girls and gossip with the old women, and we shall wait still longer. It is therefore clear that Volga himself must go.'

Lord Volga Buslavlevich turned himself into a tiny bird, and soaring up into the clouds he flew away to the land of the Turks, and coming to the Sultan's palace of white stone, he hovered against the window and listened to the secret talk.

'Ah, Queen Pantalovna,' the Turkish sultan was saying, 'do you know what I know? In Holy Russia the grass grows not as of old and the flowers bloom not as of yore. It is clear that Lord Volga Buslavlevich is alive no more.'

'Ah, royal Sultan of the Turkish land,' replied the Queen, 'I know what I know. In Holy Russia the grass grows as of old and the flowers bloom as of yore. This night I dreamt a dream. A tiny bird flew from the East and from the West a black raven, and when they clashed over the open plain and fought, the tiny bird tore the black raven all apart, plucked out his feathers and scattered them to the winds. The tiny bird was Lord Volga Buslavlevich, and the black raven was the Sultan of the Turkish land.'

'There is no truth in the dreams of old women, Queen Pantalovna!' cried the Sultan. 'I intend to march against Holy Russia! I shall take nine cities to bestow on my nine sons and shall bring myself back a fine fur coat!'

Queen Pantalovna shook her head sadly.

'You shall not take nine cities to bestow on your sons, Sultan of Turkey,' she said, 'nor shall you bring yourself back a fine fur coat.'

'Ah, you cursed old devil!' cried the Sultan. 'Take this for your stupid dream!'

And smiting her once on her white face, he smote her again on the other cheek, and threw her down on the hard brick floor a first and then a second time.

'I *shall* go to Holy Russia!' he shouted. 'I *shall* take nine cities to bestow on my nine sons, and I *shall* bring myself back a fine fur coat!'

But Volga Buslavlevich had heard everything, and turning himself into a grey wolf, he ran swiftly to the stable and tore out the throats of all the Sultan's trusty steeds. Then he turned himself into a small weasel, and scurrying to the armoury, he gnawed apart the stout bows and their silken bowstrings, snapped the well-tempered arrows in two, shattered the keen sabres and smashed the steel clubs. And when he had done all this, Volga turned himself again into a tiny bird, flew to the town of Kiev, changed back into a goodly youth and summoned his trusty druzhina.

'Brave and good druzhina,' he said, 'listen to your elder brother, your ataman, and do as I command. Follow me into the land of the Turks!'

And marching into the Turkish land, they defeated the Sultan, captured the whole of the Turkish host, and then sat down to divide the rich spoils among them.

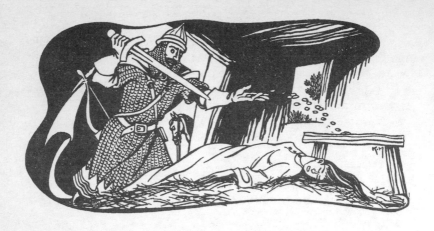

Svyatogor's Bride

THE great hero Svyatogor saddled his good steed and rode out into the open plain. As he rode, living strength flowed through his veins, but finding no one to vie with him, he felt his great strength weigh upon him like a heavy burden and he longed to put it to the test.

'If I could find a ring fixed to the earth,' he said, 'I should lift the whole world in my hand.'

Then as he looked, he saw a young man walking on foot ahead of him and he made to follow him, but though he galloped after him with all the speed he could urge from his horse, he could not overtake him.

'Hey there!' cried Svyatogor. 'You who go on foot! Wait for me, for my horse cannot catch you up!'

Looking back, the traveller stopped and waited for Svyatogor, and taking the small sack that was slung across his shoulder, he placed it down on the damp earth.

'What is in that sack of yours, friend?' asked the hero Svyatogor.

'Pick it up and see for yourself,' replied the stranger.

Dismounting from his horse, Svyatogor took the sack in one hand and pulled, and his arm was almost torn from his shoulder, for the sack seemed riveted to the ground. Then he seized the sack in two hands and pulled, and though the

8

mighty hero sank to his knees in the damp earth and the sweat and blood rolled down his face, he could raise the sack only enough for a mere breath to pass beneath it.

'What is in this sack?' gasped the hero. 'All my great strength is not sufficient to raise it from the ground.'

'The weight of the whole world is in that sack,' replied the stranger. 'It is as though there were a ring fixed to the earth.'

'Who are you?' said Svyatogor. 'What is your name?'

'My name is Mikulushka Selyaninovich.'

'Tell me, Mikulushka, how I may learn my fate as decreed by God.'

'Ride straight on until you come to a cross-road,' said Mikula. 'Take the left fork and gallop at full speed until you come to the Northern mountains. There in the hills stands a smithy under a tall tree, and the blacksmith will tell you your fate as decreed by God.'

Svyatogor left the traveller and rode straight on until he came to the cross-roads, and taking the left fork, he galloped swiftly towards the Northern mountains. His good steed sped across seas and rivers, and vast distances passed beneath its hooves, and riding for three days, Svyatogor arrived in the Northern mountains.

There in a smithy beneath a tall tree a smith was forging two thin hairs.

'What are you forging, blacksmith?' asked Svyatogor.

'I am forging the fates of those who shall wed.'

'And whom shall I wed?' asked Svyatogor.

'Svyatogor's bride dwells in the Kingdom by the Sea in the City of the King, and for thirty years she has lain on a dunghill.'

'I shall take no bride from a dunghill!' cried Svyatogor angrily, and spurring his horse, he rode swiftly to the Kingdom by the Sea, and arriving in the City of the King, he halted before a poor and lowly hut. Svyatogor looked inside, and there on a filthy dunghill, her skin as thick and black as the bark of fir-trees, lay a maiden. Taking five hundred roubles from his pocket Svyatogor laid them on the table, and taking his sharp sword he thrust it into the maiden's breast.

And when the mighty hero had ridden away, the maiden awoke and rose from the dunghill, the fir bark fell from her body, and she became a beauty such as never before or after was seen in the whole white world. And taking the money from the table, she began to trade, and when she had amassed untold golden treasure, she built herself a fleet of dark-red ships, loaded them with all manner of precious wares and sailed forth across the blue and glorious sea. When she came to the great city on the Holy Mountains, she began to sell her precious wares, and the fame of her great beauty spread through the whole land far and wide. And Svyatogor came to gaze upon the maiden's surpassing beauty, fell in love with her, wooed her, and they married.

And when Svyatogor lay beside his wife, he saw a scar on her white breast.

'What is that scar on your white breast?' he asked.

And his wife replied:

'Into the Kingdom by the Sea, into the City of the King rode a stranger, and coming to the lowly hut where I slept a deep sleep and had lain on a filthy dunghill for the space of thirty years, he placed five hundred roubles on the table. And when I awoke, there was a scar on my white breast and the fir bark had fallen from my body.'

And the mighty hero Svyatogor saw that none can escape his fate.

Dunay*

A GREAT banquet was held in the famous capital city of Kiev at the palace of courteous Prince Vladimir, and many were the princes, boyars, bogatyrs, merchants, and freemen who dined there as his guests. They ate and drank to their hearts' content, and when the sun burned red in the West and the feast grew loud and merry, Prince Vladimir rose sadly to his feet and sighed aloud.

'Alas! princes, boyars, bogatyrs, merchants, and freemen of Kiev,' said the Prince. 'All the guests at my feast are married and only I, your Prince, have no wife and must live alone. Which of you knows a princess fit for me, tall and comely and fair of face, with elegant gait and pleasing voice, with whom I might spend the rest of my days in peace?'

But all were silent and cast down their eyes, and none answered the Prince. Then the silence was broken, and bold young Dunay Ivanovich stepped from behind the oaken table. Though he had drunk much, his feet did not stumble and his tongue did not falter, and he bowed low before his Prince and beat his brow on the ground.

'Prince Vladimir, great ruler of Kiev,' he said. 'I alone, it seems, know of a princess fit to be your bride. The King

* Russian for the Danube.

11

of the good land of Lithuania whom I served for many years
has two beautiful daughters, and both are of marriageable
age. The elder daughter is Princess Nastasya, a brave
polianitsa who rides much abroad, but her younger sister
lives at home and weaves and sews indoors. Her name is
Princess Apraksia, and she has all the qualities a Prince
could desire.'

'I thank you, Dunayushka, for this good intelligence,'
said Prince Vladimir. 'Take forty thousand of my men and
ten thousand pieces of gold, go forthwith to the brave land
of Lithuania and woo Apraksia in my name.'

'Fair Sun Vladimir,' said the knight. 'I shall need
neither forty thousand men nor ten thousand pieces of gold.
Give me only my beloved comrade Dobrynia Nikitich to be
my companion and we shall bring back the fair princess to be
your bride.'

Prince Vladimir granted Dunay's request, and on the
morrow the brave knights Dunay and Dobrynia rode together out of the gates of Kiev, and those who saw them
galloping away on their fiery steeds thought they saw a pair
of falcon flying across the open plain. Such was the speed
at which they rode that they quickly reached the white
palace of the King of the valiant Lithuanian land, and dismounting beneath the window and bidding his companion
keep watch over the palace guard and come only if he should
call, Dunay went boldly in. He came before His Majesty
the King, and knowing well the etiquette of the Lithuanian
court where he had served, he did not cross himself or say a
prayer as in Kiev, but only bowed down low before the King.

'Hail, mighty King,' he cried.

'Dunay!' cried the King. 'Welcome again to our court,
brave Russian youth! You were once our trusty servant
and served us truly and faithfully for the space of three
whole years. Sit at the high table here with me and eat and
drink your fill!'

When Dunay had taken the place offered him, the King
began to question him.

'Tell me, Dunayushka,' he said, 'whither you are bound?

Have you come to visit us as a guest or once again to offer us your good services?'

'You were a good master, batiushka,' replied Dunay, 'and I served you well. But now I have not come to offer my services, but on behalf of my great master Prince Vladimir I ask for the hand of the Princess Apraksia.'

The smile faded from the King's face. He was not pleased to hear these words.

'You do wrong, Dunay Ivanovich!' he said. 'Is it right to woo the younger daughter and spurn the elder? Ho there, mighty Tatars! Seize the impudent Dunay by his white hands and hurl him into a deep dungeon! Shut him behind iron bars and oaken doors and scatter yellow sand around his cell, that he may linger awhile in the brave land of Lithuania and reconsider his unseemly proposition!'

But jumping to his sturdy feet, Dunay placed his hands on the table and vaulted nimbly across. The tables groaned beneath his weight, the green wine spilled, the dishes smashed against the floor, and the Tatars trembled in their boots and held back in fear and terror. Then suddenly more of the King's bodyguard rushed into the hall.

'Ho there, brave King of Lithuania!' they cried. 'You eat and drink and make merry and do not know the danger that menaces our land! A mighty Russian youth stands in the courtyard: in his left hand he holds the bridles of two fiery steeds and in his right hand he wields a mighty Saracen club; he springs up like the white falcon and brandishes his club in all directions. He is murdering the Tatars to the last man and will quite destroy the Tatar race!'

The King turned pale and trembled.

'Good Dunayushka Ivanovich,' he pleaded. 'Do not forget the bread and salt you have eaten in my house! Do not destroy the Tatar race, and I will gladly give you my younger daughter to be Vladimir's bride!'

And so Dunay called to Dobrynia to cease to brandish his mighty club, Princess Apraksia was prepared for the road, and Dunay and Dobrynia mounted their trusty steeds and rode from the court with the fair young princess.

After a while the dark night fell and the two brave youths dismounted and put up their tents of white linen. At their feet they tethered their fiery steeds, at their head they stuck their sharp lances in the ground, at their right hand they placed their keen sabres and at their left hand their steel daggers, and then they lay down to sleep. The two brave youths slept on until the night grew short, and then they heard the beating hooves of a pursuer drumming across the open plain. The next morning they rose early and went their way, but still the Tatar pursued them, his horse sinking up to its fetlock in the mud and scattering the pebbles for two bowshots all around. And sending Dobrynia on ahead with Princess Apraksia, Dunay Ivanovich wheeled his horse round and rode to meet the Tatar.

The Tatar roared like a wild beast and hissed like a serpent, and bore down upon the valiant Russian knight. The dark woods shook, stones flew up, the green grass faded and the flowers scattered, but Dunay leapt nimbly from his horse, and standing firmly on his sturdy legs, he smote the Tatar with his mighty fist as he passed, knocked him from his horse into the dust, and knelt upon his armoured breast.

'Before you die, Tatar dog,' he said, 'tell me who you are, of what race and of what family?'

'Ach, Dunay Ivanovich!' cried the Tatar. 'Were I kneeling on your breast I should not ask your name or family or race, but should rip up your white breast with my keen sword!'

Angered by these words, Dunay tore away the Tatar's cloak and made to plunge in his knife and rip up the Tatar's breast. But suddenly his hand faltered and his heart was filled with mercy. His vanquished enemy was a woman!

'Did you not recognize me, Dunayushka?' she said. 'There was a time we sat at the same table, ate from the same bowl, and rode together across the open plain. Did you not serve my father for the space of three years?'

And when he recognized the elder daughter of the King of Lithuania, Dunay thanked God that he had stayed his hand.

'Come, brave Nastasya!' cried Dunay Ivanovich. 'Let us journey quickly to the town of Kiev, where you and I shall stand beneath the golden crowns of marriage.'

And so the younger sister Apraksia and the elder sister Nastasya married in the same church, the one to Prince Vladimir of Kiev and the other to the bold knight Dunay Ivanovich, and when the marriage feast was in its third day, Dunay rose to his feet and began to boast.

'In all the famous city of Kiev,' he cried, 'there is none to compare with Dunay Ivanovich! He brought back a wife for his prince, Fair Sun Vladimir, and he brought back a wife for himself.'

'Ach, Dunayushka!' cried Princess Nastasya. 'Your proud boasts are hollow and empty. I have not been long in the city of Kiev, but I have learned not a little of its men. I know there is none to equal Churilo Plenkovich in finery and none to equal Alyosha Popovich in valour, and now there is none to equal me, Princess Nastasya, in archery. If you doubt me, set up a knife at a great distance and I will fire my arrow so that it shall split in two against the blade of the knife, and the two halves will be equal in size and weight.'

'Well, at least you have learned to boast like a Russian, wife,' laughed Dunay Ivanovich. 'Let us go into the open plain, and we shall see who is the better archer of the two!'

A sharp knife was stuck blade-upward in the ground in the open plain and Nastasya took her bow and shot her well-tempered arrow from the silken string. And striking against the blade of the knife, the arrow was cut into two halves equal in size and weight. Then Dunay took his bow and shot his arrow, but his aim was high and the arrow fell beyond the knife. He shot a second arrow, but his aim was low and it struck the ground short of the knife. And though he shot a third arrow still he could not hit the knife, and hearing Nastasya's laughter in his ears, Dunay Ivanovich grew angry, and fitting another arrow in his stout bow, he took aim at Nastasya's white breast.

'Dunayushka!' cried Nastasya. 'Do not shoot your arrow at my breast! Soon I am to bear you a son: his legs are silver to the knees, his arms are gold to the elbows, stars twinkle in his hair, and he shines through the darkness like the red sun itself.'

But hot anger deafened Dunay's ears and hardened his brave heart, and unheeding his fair wife's pleas, he loosed his well-tempered arrow into her white breast, and she sank dying to the ground. But when Dunay came to his senses and saw that he had killed his wife Nastasya and that his son would never be born, he beat his breast and tore his hair in grief, and driving the hilt of his sword into the damp earth, he fell upon its point and died.

And where the head of his fair wife fell to the ground, the River Nastasya sprang forth; and where the head of brave Dunay fell, sprang forth the great River Danube.

Dobrynia and Alyosha

THE good knight Dobrynia sat at his window in the famous city of Kiev and gazed out into the open plain, and seeing him brooding thus, the old widow Omelfa Timofeyevna came to his side and said:

'Is it not time, Dobryniushka, that you should take a wife?'

'Gladly would I marry, Mother,' replied her son, 'but are not all the fair maidens of Kiev already wed? To find a maiden to be my bride, I should have to journey to distant cities, to far-off villages and towns.'

Dobrynia stood up on his sturdy legs and left the window.

'But it is wearisome to sit for ever inactive here,' he cried. 'I shall journey out into the open plain, down to the Puchay River.'

'Go then, Dobrynia Nikitich,' said the widow Omelfa Timofeyevna. 'Take not your own horse Voroneyushka, Little Raven, but take the steed of your dead father; take, too, his keen sword and mighty club and silken whip, for your father never rode abroad without his whip.'

Dobrynia took his father's horse and sword and club and whip and rode out into the open plain. Before long he came upon the tracks of another rider, and spurring his horse on in pursuit, he drew near to a valiant polianitsa riding her jet black horse across the open plain, and taking his strong bow, the Russian knight placed an arrow of tempered steel on its silken string.

'Fly, my arrow, fly,' he cried, 'and strike the female warrior on her turbulent head!'

The arrow shot from the bow and struck the woman on her head. But the arrow fell harmlessly to the ground and it was as though she felt nothing, for she rode calmly on and did not look back.

Dobrynia took a second arrow and shot it from his sturdy bow, and again it struck the warrior on her turbulent head, but noticing nothing, she rode calmly on and did not look back.

A third time Dobrynia's silken bowstring twanged and a well-tempered arrow struck hard against the polianitsa's head, and this time she looked back and laughed aloud.

'I see now,' she cried, 'that it is a Russian warrior that shoots his arrows against my head! And I thought it was the feeble Russian gnats!'

And reaching down from her horse, she seized Dobrynia by his golden curls, and thrusting him deep into her wide pocket, for three days and nights she carried him thus.

On the fourth day, however, her trusty steed stumbled and called out aloud in an anguished voice:

'Alas, young mistress, sweet Nastasya Nikulichna! Before I bore only you, fair maiden, valiant polianitsa, but now I bear upon my back a mighty Russian knight as well. It is grievous hard for me to carry two great bogatyrs!'

And Nastasya was eager to know the name of the prisoner her horse had called a mighty Russian knight, and reaching down into her pocket, she brought him into the light and questioned him.

'What manner of knight are you, then?' she asked. 'Of what land, of what town and village, of what mother and what father were you born? Are you the son of a prince, the son of a king, the son of a merchant, or the son of a simple peasant?'

'What is that to you, polianitsa?' cried Dobrynia angrily.

'I shall tell you neither of my father and my mother, nor of what town and village I may be!'

'Ach, impudent Russian!' laughed Nastasya. 'If I knew you were old, I would strike off your head without more ado; if I knew you were young, I would call you my blood-brother; but if I knew you were equal in years to me, I should take you to be my husband!'

'He is a mighty Russian bogatyr!' said her horse. 'His name is Dobrynia Nikitich. In strength he is comparable with you, in valour he is as two of you, and in years he is your equal.'

Hearing these words, Nastasya Nikulichna released Dobrynia Nikitich and they rode along together, each on his own horse.

'Marry me now, Dobryniushka,' said Nastasya, and riding together into Kiev, they entered the spacious courts of Dobrynia's residence, and went in at the door of the palace of white stone.

'I have found her who is destined to be my bride, Mother,' said Dobrynia to the widow Omelfa Timofeyevna. 'I shall lead Nastasya Nikulichna to the Church of God, and there they shall hold the golden crowns of marriage over our turbulent heads.'

When they were married, a great banquet was held. Prince Vladimir came, and Princess Apraksia and all the Russian bogatyrs, and for three days and for three nights the feast went on. When the wedding guests had returned to their homes, Dobrynia remained alone with his young wife Nastasya Nikulichna, and she ceased to ride out into the open plain, and remained like any other wife at home with her husband's mother, the widow Omelfa Timofeyevna. And thus they lived for the space of three years.

Then when the three years had passed, Prince Vladimir of Kiev arranged another great banquet and invited Dobrynia to the feast, and when the day had passed in feasting and the banquet drew to a close, Prince Vladimir stood up and said:

'Which of you mighty Russian bogatyrs will go to the

aid of my father-in-law the King of Lithuania and fight the army that has attacked his land, for he has written to ask me for the help of a Russian knight? Which one of you will perform this service for my sake?'

All the bogatyrs were silent and looked down, and none answered the Prince.

Then the Prince asked a second time and received no answer.

Then the Prince asked a third time, and from the white oak bench up rose the mighty Russian bogatyr, the young Alyosha Popovich.

'Prince Vladimir, Fair Sun,' he said. 'Apart from Dobrynia no one of us has ever seen the valiant Lithuanian land. But Dobrynia once served its King and will know how to fight after their own fashion.'

Then the Prince said to Dobrynia:

'If you will not carry out my command, Dobrynia, you will incur my great wrath and I shall banish you from the famous city of Kiev for ever! But if you carry out my command, I shall be exceedingly grateful and shall honour you well.'

Dobrynia stood on his sturdy legs, and grimly thanking Prince Vladimir and Princess Apraksia for the honour they had bestowed upon him, he marched out of the hall and returned home, and sadly began to prepare for his journey to the distant land of Lithuania. Neither to his wife nor to his mother did he breathe a word of his mission, but the handmaids and servants quickly learned of it, and ran to tell the bad news to Omelfa Timofeyevna his mother. And beating her breast, she ran in tears to her son.

'Where are you going, my child, my precious pearl, my sweet berry?' she wept. 'Why do you say nothing to your mother and keep your secret from your wife?'

'Why do you so question me, Mother?' cried Dobrynia. 'If I live, I shall return, and if I die, do not wait for me. My heart is wroth against you, Mother, for you it was gave birth to me, Dobryniushka, unhappy son of Nikita. Where is my erstwhile happiness fled, and for whom has it

abandoned me? Better it were by far had I never been born
into this white world! Were I but a grey stone on the bed of
the Puchay River, a thousand years should I lie there undisturbed in peace, and mighty Russian heroes riding by
should marvel at that stone. Were I a beech-tree in the open
plain, these same heroes would rest in my leafy shade. Had
you but laid me at my birth in a basket on the waves of the
Dnieper River, the swift stream would have borne me off,
and never again should I have suffered in the wide, white
world!'

In tears the good widow Omelfa said:

'Gladly would I have brought a happier child into the
white world, a son with the luck of Ilya of Murom, the
strength of Svyatogor, the gentleness of David, the wisdom
of Solomon, the valour of Alexander, the beauty of Joseph,
the wealth of Sadko of Novgorod, the grace of Churilo
Plenkovich, the courtesy of Dobrynia Nikitich, the hand-
writing of Dunay Ivanovich, and the daring of Alyosha
Popovich! But alas, Dobryniushka, I have no other sons
but you, and if God has not already given you good fortune,
He shall not do so now.'

And weeping bitterly, she ran quickly to the white-stone
palace to her God-given daughter-in-law, young Nastasya
Nikulichna.

'Alas, Nastasya!' she cried. 'You sit idly at the window
and do not know the peril that hangs over you! Your
husband is donning his knightly armour and prepares to
leave you. Run, stand by his right stirrup, and say with
a smile that which I teach you now to say. Speak thus to
your husband before he leaves: "O my dearest, my darling,
did you not once take me for your wife? Did we not stand
together in the Church of God with the golden marriage
crowns above our heads? For three years we have lived in
good accord, and it is sad for me to see you go. Speak, beloved husband, pleasant words to me, and tell me whither
you are bound, and whether your journey will be long or
short?" '

And Nastasya ran to her husband and stood by his right

stirrup, and spoke the words the widow Omelfa bade her say.

'Alas, dear wife,' said Dobrynia. 'If I live, I shall return, and if I die, do not wait for me.'

But Nastasya asked him a second time, and a third time, and Dobrynia said:

'If I do not return in three years, wait for me another three. Perform your duties as a wife, and sit and wait for my return. If I return not then, wait for me another six years, and when these twelve years have passed, perform the duties of a widow, give my golden treasure to the Church and monasteries in my memory, to the poor and the widowed and the little orphans. And you, Nastasya, shall be free to live as a widow or to marry again. Marry a prince or the son of a prince, a king or a Russian knight, a rich merchant or a simple peasant, only do not marry Alyosha Popovich, for we are brothers by exchange of crosses.'

And whipping the flanks of his horse, Dobrynia galloped out of the wide courtyard into the open plain, and coming to the land of the infidel, began to fight against the enemy host.

Long years passed in battle, and still Dobrynia Nikitich did not return home, and when three years had gone by, the brave knight Alyosha Popovich rode into Kiev from the open plain.

'Great Prince, Prince Vladimir,' he said. 'A dead knight lies in the open plain. His turbulent head is sundered from his body and lies beneath the willow-tree. The green grass grows through his yellow curls, and by the flowered coat I knew it was Dobrynia Nikitich, my brother in Christ. Go now to his mother, Fair Sun Vladimir, and tell her the grievous tidings I bring.'

Prince Vladimir put on his robes and went quickly to Dobrynia's mother, the widow Omelfa Timofeyevna, and when she heard the news, she burst into tears and wept bitterly.

But Nastasya Nikulichna shook her head.

'I do not believe the words of Alyosha Popovich,' she said,

'for he does not speak the truth. I shall wait another three years for my husband Dobrynia.'

But when six years had passed and Dobrynia had not returned, Alyosha rode in again from the open plain and brought news that Dobrynia had been killed in the land of the infidel. And again the Prince went to Dobrynia's mother, the widow Omelfa Timofeyevna, and told her these sad tidings. But although Omelfa Timofeyevna wept, Nastasya Nikulichna only shook her head and said:

'What Alyosha Popovich says, I do not believe. I shall wait for my husband.'

When another six years passed by, now twelve in all, and Dobrynia had not returned to Kiev, Alyosha Popovich bade Prince Vladimir go to the widow Omelfa Timofeyevna and ask Nastasya Nikulichna to be his bride, for her husband, he said, was certainly no more among the living. The Prince went to Dobrynia's white-stone palace, and the widow came to meet him.

'Greetings, Fair Sun, Prince Vladimir,' she said. 'What good tidings do you bring?'

'Alas, I bring no good tidings, Omelfa Timofeyevna,' said the Prince. 'Your son Dobrynia is dead, and his wife Nastasya, who has waited faithfully for him for twelve long years, must wait no longer. Give Nastasya to be Alyosha's bride.'

'I cannot speak for Nastasya,' replied the widow. 'If she wishes to marry Alyosha, let her marry him. But if she does not wish to marry Alyosha, I shall not force her against her will.'

'I waited six years for Dobrynia,' said Nastasya, when she heard of the Prince's mission, 'and I did my duty as a wife. Another six years have passed and Dobrynia has not returned, and I have done my duty as a widow. I shall wait for him no more. I shall marry Alyosha Popovich.'

And thus it came about that arrangements were made to celebrate the wedding of Alyosha Popovich and Nastasya Nikulichna, and the people of Kiev looked forward to the great feast.

And all the time, far away in the open plain in the valiant Lithuanian land, Dobrynia Nikitich slept the deep sleep of a Russian hero in his tent of white linen; and just as he was about to open his eyes, two white doves flew up, and one of them settled on top of Dobrynia's tent.

'Do not stay to rest now, brother!' exclaimed the other. 'Come and fly with me to the stone-walled city of Kiev. Twelve years have passed since Dobrynia rode out into the open plain and left his young wife Nastasya alone in Kiev, and now she has taken another husband—not a prince or a boyar's son, not a mighty Russian bogatyr, but the young Alyosha Popovich, Dobrynia's brother by exchange of crosses.'

When Dobrynia heard this, he leaped to his feet, jumped on his trusty steed Voroneyushka, Little Raven, lashed it with his silken whip, and galloped furiously away to Kiev. When he arrived in the wide courtyard of his home, he did not tie his horse to the golden ring on the post, but riding straight to the stables, he dismounted and walked boldly into the white-stone palace.

The servants and handmaids followed him in great distress, and complained to Omelfa Timofeyevna that a ruffian had ridden in and treated the house as it were his own. Dobrynia's mother buried her head in her hands and wept.

'Alas, alas!' she cried to the intruder. 'If my brave son Dobrynia were still alive, you should not so insult my widowed and orphaned house.'

'Do not be angry with me, Omelfa Timofeyevna,' said Dobrynia. 'I have been sent by your son Dobrynia, for only yesterday did I meet him as he rode across the open plain, and he bade me come to you and bring you his humble greeting.'

'You mock me, mighty Russian bogatyr!' cried the widow. 'If my son were among the living, he would have come to greet me in person.'

Dobrynia could bear his mother's distress no longer.

'I am surprised at you, Mother!' he cried. 'Can it be that you no longer recognize your son Dobrynia?'

But Omelfa Timofeyevna gazed at him with no sign of recognition in her eyes.

'You mock me again,' she said. 'My Dobrynia was young, his complexion was white as the powdered snow, and his clothes were adorned all with flowers!'

'Twelve long years have passed, Mother,' answered Dobrynia. 'The rain has washed the flowers from my coat, and the red sun has darkened my once white face.'

Omelfa Timofeyevna faltered and said:

'If you are my son Dobrynia, show me the birthmark on your right knee.'

And throwing off his boots of moroccan leather, the hero revealed his birthmark to his mother, and she knew her long-lost son, and wept.

'Alas, Dobrynia!' she cried. 'They brought me rumours of your death that broke my heart and filled my bright eyes with tears! My son, you come too late! For six years Nastasya fulfilled her duties as a faithful wife, and for six years she fulfilled her duties as a faithful widow, and now she has taken Alyosha Popovich to be her husband.'

'Bring me my flowered garments, Mother,' cried Dobrynia, 'and my maple psaltery. I shall dress myself as a minstrel and go thus to the marriage feast.'

And so Dobrynia went as a minstrel to the palace where the feast was held, and crossing himself and bowing low before Prince Vladimir and Princess Apraksia, he said:

'Greetings, Prince Vladimir of Kiev. I am a minstrel from foreign parts. Pray designate the place where I shall sit.'

'You place is that of the minstrel, stranger,' said Vladimir. 'Sit on the glazed tile stove.'

Dobrynia sat in the place the Prince had designated, and taking his maplewood psaltery in his hand, he began to sing sweet songs from Tsargrad; and he sang too of the fair city of Kiev and its great Prince Vladimir, and songs in honour of the marriage of Alyosha and Nastasya.

Nastasya Nikulichna was charmed by his sweet voice and gracious manner, and turning to Prince Vladimir, she said:

'Permit me, Great Prince, to pour out a cup of sweet mead and to give it to whomsoever I will.'

'Pour out your cup of sweet mead, Nastasya Nikulichna,' said Vladimir, 'and give it to whomsoever you will.'

And pouring out the cup of mead, Nastasya gave it to the minstrel. Dobrynia took it in one hand and quaffed it at a draught, and turning himself to the Prince, he sought his permission to offer a cup of sweet mead to whomsoever he chose. And when the permission was granted, Dobrynia poured out a cup of mead, and secretly slipping his golden wedding-ring into it, he gave the cup to the fair Nastasya. And when Nastasya drank, the golden ring slipped against her lips, and she straightway knew the minstrel to be her husband Dobrynia. She clutched the ring in her left hand, and resting her right hand on the white oak table, she sprang across and fell at the feet of Dobrynia.

'Forgive me, Dobryniushka!' she cried.

'I forgive you, Nastasya Nikulichna,' said Dobrynia. 'A woman's hair is long, but her wit is short, and does not the proverb say that man has only to go into the forest to fetch wood for his wife to marry another? I forgive you, Nastasya, for it is not at your wit that I wonder, but at that of Prince Vladimir, who gives the wife of one man to another!'

The vast hall was silent, and Prince Vladimir, and all the princes and boyars and merchants hung their heads, and Alyosha Popovich turned pale and trembled.

'Forgive me, Christian brother, forgive me,' he cried. 'I have taken your beloved wife Nastasya and am guilty of mortal sin.'

'For that fault, brother,' cried Dobrynia, 'it is God who must forgive you. But that you said I was dead and hurt my mother and made her weep, for that I shall never forgive you!'

And leaping at Alyosha, Dobrynia seized him by his golden hair and threw him roughly down upon the tiled floor; but as he made to strike off his turbulent head with his keen sword, the old Cossack, Ilya of Murom, held back his hand.

'Does the proverb not say that all may marry but few are happy?' he said. 'Part now, brothers, and be reconciled.'

And Dobrynia listened to his words, and leaving Alyosha Popovich lying on the floor, he took his fair wife Nastasya back to his mother, the widow Omelfa Timofeyevna, and thus, after twelve long years, were they all united.

Ilya of Murom and Svyatogor

NEAR the famous town of Murom, in the village of Karacharov, a child was born to worthy and honourable parents who called him Ilya Ivanovich, by surname Muromets. But the legs of their son had no strength to walk and his hands had no power to hold, and for thirty years he remained indoors, helpless and idle, a cripple.

One summer when his mother and father had gone into the fields to work and Ilya remained alone, three wayfarers much advanced in years passed by the little window.

'Hail, Ilya of Murom, peasant's son!' they said. 'Open your doors to three wandering pilgrims!'

And Ilya did not know that they were Christ our Lord and two apostles, and he sighed aloud and said:

'Greetings, wandering pilgrims! Gladly would I help you, but alas! I cannot open the doors to let you in, for I have sat thus for thirty years and have no power over my arms and legs.'

But the wandering pilgrims said:

'Stand up, Ilya, on your sturdy legs, open the door with your mighty hand, and give us to eat and drink, for we hunger and thirst from much travelling.'

And straightway Ilya stood up upon his sturdy legs and looked up at the holy icon.

'Glory to Thee, O God!' he cried. 'The Lord God hath given me strength in my legs and power in my arms.'

And opening the door for the wandering pilgrims, he went down into the deep cellars and brought back a full cup of mead and bade the ancient men drink thereof. And when they had drunk, they said to Ilya:

'Go down again into the deep cellars, Ilya of Murom, and bring yourself a full cup of mead to drink to your own health.'

And Ilya went down again into the deep cellars, and bringing back a full cup of mead, he drank, and his heroic heart burned and his white body glowed.

'What do you feel within you, Ilya?' asked the three pilgrims.

'I feel a great power move within me, and if there were a ring set in the damp earth, I could turn the whole world on its side!'

'Then go down again into the deep cellars,' said the old men, 'pour yourself another glass of mead and drink.'

And when Ilya had done as he was bidden, the old men asked him what he now felt within him.

'I feel that half my strength has drained from me,' replied Ilya.

'So be it, Ilya,' said the pilgrims. 'Thus shall you live. You shall be a mighty warrior. Death shall not come to you in the open plain, nor in any battle, and you may fight and vie with any bogatyr. But do not contend with the mighty hero Svyatogor, for the Earth can barely hold him up; do not fight with Mikula, for he is beloved of Damp Mother Earth; fight not with Volga Yaroslavich, for though he will never take you by force, his artful cunning will defeat you. You must get yourself a powerful warhorse, Ilya, and this is what you must do: ride across the open plain and purchase the first foal you see. Keep the foal for three months in the stable and feed it on the whitest wheat, and when the three months are past, tether the foal in a garden for three nights and sprinkle it with three morning dews. Then lead the foal to a high hedge, where the foal will begin

to jump backwards and forwards across the hedge. And
when you have done all this, ride your horse whithersoever
you will, for it shall bear you.'

And with that, the three pilgrims vanished from Ilya's
sight.

Thus endowed with new heroic strength, Ilya went to find
his parents to help them with their work, and when he came
upon them, they were fast asleep, reposing from their toil
of felling the sturdy oaks. Leaving his parents undisturbed,
Ilya tore out the massive oak-trees by the roots, threw them
into the deep Dnieper river until its mighty course was
almost blocked, and returned home; and when his father and
mother awoke from their deep sleep and saw what had been
done, they were afraid.

'Who has done this miracle?' they said, and since their
work was done, they, too, returned home.

When they entered their hut, they saw their son walking
on his sturdy legs, and when they had recovered from their
surprise, they asked him what had happened to make him
well. Ilya told them of the three strange pilgrims and how
they had bidden him drink a cup of mead, and how new
power had come into his arms and legs and how his body
had filled with strength.

'God has given you great strength, Ilya,' said his parents.
'Live now in modesty and humility, and do not give way to
the promptings of your stout heart.'

Ilya walked out into the open plain, and seeing a peasant
leading a mettlesome foal, grey-brown and shaggy-haired,
he bought it, and returning home, kept it in the stables for
three months and fed it on the whitest wheat. When the
three months were past, he enclosed it in a garden for three
nights and besprinkled it with three morning dews, and
when he led it to a high hedge, the horse began to jump from
one side to the other. And Ilya named the horse Sivushko,
and putting a saddle and bridle upon it, he led it before his
parents.

'Father and mother,' said Ilya, 'give me now your blessing, for I am going to the city of Kiev, to the Fair Sun Prince

Vladimir, to pray in the holy churches to the Holy Mother of God!'

And though his father and mother tried to persuade him to stay, Ilya insisted, and they gave him their blessing.

'Go my beloved son, to the famous city of Kiev,' said his mother, 'but do not stain your sword with blood, nor orphan little children, nor widow young wives.'

And bidding farewell, Ilya mounted his steed, and beating it on its sloping flanks and between the ears, he galloped away, and Sivushko flew like a falcon over rivers and lakes and swept the fields with its tail.

And all the elder bogatyrs marvelled to see it and said:

'None rides so well as Ilya of Murom, for he sits like a hero, and his bearing is that of a fine bogatyr!'

As Ilya of Murom journeyed through the open plain he came upon a huge pavilion of white linen pitched beneath an oak-tree, and dismounting from his horse, he entered boldly in. Inside he saw a hero's mighty bed of iron, and Ilya marvelled greatly at it, for it was fully ten fathoms long and six fathoms wide.

'What manner of man sleeps in this bed?' he wondered. 'Can it be an infidel and evil Tatar or a powerful Russian bogatyr?'

And tying his good steed Sivushko to the oak-tree, Ilya lay upon the huge bed and went to sleep, and since the sleep of the bogatyr is long and deep, he slept thus for three days and three nights. On the third day his horse heard a great noise approaching from the North. Mother Earth trembled, the dark forests shook, rivers overflowed their banks, and Sivushko stamped the ground with its hoof, but could not wake Ilya his master.

'Ilya of Murom!' cried the horse in a human voice. 'You sleep and take your ease and know not what disaster looms above you. The mighty hero Svyatogor is returning to this his tent! Let me loose in the open plain and climb yourself up into the oak tree.'

Remembering his pilgrims' warning, Ilya leapt to his sturdy feet, let his horse loose in the open plain and climbed quickly into the leafy oak-tree. And then he saw Svyatogor approach, and the hero towered so high above the standing woods that his head reached up to the drifting clouds. On his shoulder he bore a crystal casket, and when he came beneath the tree, he took the casket from his shoulder and opened it with a key of gold. Out of the casket stepped Svyatogor's heroic wife, and never in the whole white world was her like ever seen. Tall in stature and graceful of gait, with eyes keen as the hawk's and brows as black as sable, she stepped out of the crystal casket and began to cover the oaken table with all manner of sweet meats and honey drinks, and then she and Svyatogor sat down to eat. When they had eaten, they played a game of cards until the mighty hero felt drowsy and went into the tent to sleep. Hereupon his young wife prepared to walk in the open plain, but as she passed beneath the oak-tree, she looked up and saw Ilya of Murom hiding in the foliage.

'Ho there, young hero!' she cried. 'Come down from the oak and dally with me! If you will not obey me and do what I ask of you, I shall awake the hero Svyatogor!'

Ilya came down from the tall oak and did as he was bidden, and taking Ilya and putting him into her husband's deep pocket, the faithless woman went to awake the hero Svyatogor. The great bogatyr rose, placed his fair wife into the crystal casket, mounted his trusty steed, and rode towards the Holy Mountain. Soon, however, his horse began to stumble, and when Svyatogor began to beat it with his silken whip, it cried out in a human voice:

'Once I bore a hero and his heroic wife upon my back, but now I bear a heroic wife and two mighty heroes. Is it a wonder I stumble?'

Thereupon Svyatogor realized his fair wife's treachery, and taking the silver casket from his shoulder, he took her in his hand, and seizing his keen sword, cut off her turbulent head. Then Svyatogor took Ilya from his pocket and began to question him.

'What is your name, good youth, and whither are you bound?'

'My name is Ilya, from the town of Murom, from the village of Karacharov,' replied the hero. 'And I desired to behold the hero Svyatogor, for I have heard that he no longer rides on Damp Mother Earth and no more appears among the bold heroes of Russia.'

'I would willingly ride among you,' replied Svyatogor, 'but Mother Earth could not carry me. I cannot ride in Holy Russia, but only among lofty hills and rocky ravines. Come, let us ride together to the Holy Mountains.'

And as they rode through the mountain passes, Svyatogor taught Ilya all the heroic holds and stances, and when they had ridden some way, they came to a large coffin lying on the road, and they marvelled at it and wondered for whom the coffin was made.

'Lie in it, Christian brother,' said Svyatogor. 'Perhaps it may fit you.'

But when Ilya stepped into the iron coffin and lay down, it was both too long and too wide, and he seemed as small as any babe therein.

Then Svyatogor himself lay down in the iron coffin, and it fitted his body perfectly, and was neither too large nor too small.

'This coffin is made for me,' laughed Svyatogor. 'Take the coffin-lid and put it over me.'

'You have jested enough, brother,' cried Ilya. 'I shall not take the coffin-lid and cover you up. Do you wish to bury yourself for ever?'

Svyatogor laughed again, and taking the coffin-lid in his own hands, he placed it over him on the coffin. But when he tried to get out again, he could not rise, and though he kicked and strained with all his might, he could not move an inch.

'Alas, younger brother,' he cried to Ilya of Murom. 'Fate has sought me out. Help me, try to raise the coffin-lid, for I cannot.'

But however hard Ilya struggled with the lid of the iron coffin, he could not shift it even a little.

'Take my keen sword, Ilya,' said Svyatogor, 'and strike across the coffin-lid.'

But when Ilya grasped the hilt of Svyatogor's great sword, he could not lift it from the ground.

'Bend over the join in the coffin, Ilya,' cried Svyatogor, 'and I shall breathe into you the strength of heroes.'

Ilya bent over the coffin and Svyatogor breathed upon him, and Ilya was filled with heroic power thrice greater than before. And picking up the great sword, he struck the coffin a fearful blow across the lid. Blue sparks flashed and a ribbon of iron sprang off, but the coffin-lid stayed on.

'I am stifling, younger brother!' gasped Svyatogor. 'Bend down again and I shall breathe into you a strength greater than before. Then strike a blow along the lid of the coffin.'

Ilya bent over the coffin, and when Svyatogor breathed upon him, he was filled with heroic strength thrice greater than before, and he smote the coffin a fearful blow along the lid. Blue sparks flashed and a ribbon of iron sprang off, but the coffin did not break.

'Young brother, I am dying!' cried Svyatogor. 'Bend down again, and I shall breathe into you all my great heroic strength.'

'My present strength suffices me,' said Ilya. 'Had I more power, the Earth could not carry me.'

'You do well not to heed my last command, younger

brother,' said Svyatogor, 'for I should have breathed into you the spirit of Death, and you would have fallen dead beside me. Take now my mighty sword, but tie my good steed to my coffin, for no one but me may ride that horse. Ride into Holy Russia, mighty hero, for though you contend with all the bogatyrs, none shall slay you. You were not born to die in the open plain, Ilya of Murom, but in your native home. And now, farewell for ever!'

Svyatogor's last dying breath rose from the coffin. Ilya bade farewell to the mighty hero, and tethering his trusty steed to the coffin, he girt the hero's mighty sword about his loins, and rode back into the open plain.

Ilya of Murom and Nightingale the Robber

THE great Russian hero Ilya of Murom, son of a peasant, saddled his horse and rode forth in search of adventure and heroic exploit. And as he rode, he beat his horse's flanks so hard with his silken whip that it bit through the thick hide into the black flesh beneath, and the horse reared up and leapt from the ground, and bounding high above the standing trees, it soared beneath the drifting clouds. The first leap was of fifty versts. Then the horse leapt again, and where it came to earth a well appeared, and striking down a tall oak-tree, Ilya built a little chapel beside the well and wrote upon it:

'Here rode the mighty and powerful bogatyr
Ilya son of Ivan, Ilya of Murom.'

Then the horse leapt a third time and alighted within sight of the city of Smolensk, and Ilya saw that the city was besieged by an immeasurable Tatar horde, for three infidel Tsareviches stood there, each with an army of forty thousand men. The Tatar hosts were as thick as a swarm of black arrows: no bird could fly over them, no beast run through them, and no man pass by them. Ilya's heart leapt in his breast, and seizing his mighty sabre, he fell on the Tatar host and brandished it aloft. Where he struck, a road appeared, and where he parried, a lane. Tatars fell to the right, and Tatars fell to the left, and Ilya ploughed through their bodies towards the three princes in the middle of the field.

'Ho there, you sons of kings,' he cried. 'Shall I take you prisoner or strike off your turbulent heads with my keen sword? I have a journey to make and nowhere to put three prisoners, but if I strike off your heads, a race of kings shall perish. Go back to your lands and tell the world that Holy Russia does not lie at the mercy of the likes of you and still has mighty heroes to defend her glory!'

And speaking thus, Ilya turned his back and rode on into the city of Smolensk. But no man greeted him as he rode through the gates. All the streets were deserted, for all the townsfolk had prepared to sally forth to certain death against the Tatar foe, and were now, great and small, all assembled in the Church to pray and repent and take their leave of the white world. Ilya of Murom rode to the Church, tied his horse to the golden ring in the carven pillar, and entered. He crossed himself, bowed down on all sides, and then spoke these words:

'Ho there, men of Smolensk! Why do you repent already of your sins and prepare for Death?'

'Stranger, repent also!' they answered. 'Before our city of Smolensk stands an untold host of pagan ulans and Turkish mirzas, and they have threatened to slay and plunder and cut off our heads and make us captive.'

'You repent too soon, men of Smolensk!' cried Ilya of Murom. 'Climb on to the city wall and look out into the open plain. Tell me, do the Tatar hosts stand or lie, do they wake or sleep?'

And when they looked out from the city walls, they saw all the pagan Tatars lying dead upon the ground, and they fell on each other's necks and wept for joy.

'Brave youth,' they cried, running down from the city walls. 'You have slain our deadliest enemies. Live with us here and be the governor of the city of Smolensk!'

'God forbid that a peasant become a noble, a noble a peasant, a priest a hangman, or a warrior a governor!' replied Ilya, mounting his trusty steed Sivushko. 'Tell me rather the shortest road to Kiev.'

'Ilya of Murom,' said the citizens of Smolensk. 'The straight road to Kiev is closed, for it leads through the Black Mire, which sucketh all to certain death, and across the broad and tumultuous River Smorodina. In the Brianski Woods a band of robbers lie in wait, and for thirty years the Robber Nightingale has sat on the tops of seven oaks. He pipes like a nightingale, hisses like a serpent, roars like a bull, and shrieks like a Tatar, and all who hear his voice fall dead on the spot! Take rather the long road to Kiev.'

'I shall take the straight road,' said Ilya, and whipping his horse, he galloped away towards Kiev.

Sivushko leapt safely across the Black Mire and soared over the raging torrents of the River Smorodina, and Ilya rode boldly into Brianski Wood, where the Robber Nightingale, son of Rakhmat, sat on the tops of seven oaks. And when he saw the Russian hero come, Solovey Rakhmatich piped like a nightingale, hissed like a serpent, roared like a bull, and shrieked like a Tatar. The green grass flattened, the blue flowers scattered, the dark forest bowed to the ground, and Ilya's good horse Sivushko fell to its knees.

'Get up, you wolves' meat, you bag of hay!' cried Ilya, kicking his horse in the belly. 'Have you forgotten how to walk, you dog? You have heard before the pipe of the nightingale, the hiss of the serpent, the roar of the bull, and the shriek of the Tatar! Why, then, do you stumble now?'

But the horse could not rise, and taking his mighty bow, Ilya placed therein an arrow of tempered steel and drew the silken bowstring to his shoulder.

'Fly, my arrow!' he cried. 'Fly over the standing wood,

soar beneath the drifting cloud, strike Nightingale in his right eye and come out through his left ear!'

The silken cord twanged, the well-tempered arrow flew high above the tall woods and soared beneath the drifting clouds, and striking the robber in his right eye, it came out through his left ear, and Nightingale Rakhmatich fell from his nest like a mighty rick of hay.

Ilya seized Nightingale by his golden curls and tied him to his left stirrup, and leading his horse with his left hand, Ilya struck down tree after tree with his right and advanced thus through the thick forest until he came to the robber's white palace.

'Look, daughters!' cried Nightingale's wife, when she saw the group approaching. 'Your father is bringing us a sturdy Russian youth.'

When the robber's daughters looked out of the window, however, they saw that it was not so.

'Our father Nightingale leads nobody, Mother,' they said, 'but is himself lead hither in ignominy. His eye is pierced by an arrow and he is tied to the youth's steel stirrup by his golden curls.'

'Open the gate!' cried the mother. 'Fall upon the Russian hero with your horns of wild beast, tear him to pieces, and eat him up!'

But Nightingale Rakhmatich heard his wife's voice.

'Throw away your horns of wild beast, my children,' he cried, 'and welcome Ilya, son of a peasant, to our dwelling. Give him sugared viands to eat and sweet mead to drink, for he is too powerful a foe for you!'

But the robber's eldest daughter paid no heed to her father's words, and leaping out into the wide courtyard, she seized the cast-iron threshold plate weighing ninety poods, and swinging it high in the air, she made to smite Ilya between his two clear eyes. Ilya dodged the mighty blow, but it glanced off and numbed his right arm; and drawing back his foot, the hero kicked Nightingale's eldest daughter under the rump, and soaring high above the white palace, she fell on her back by the garden wall.

'Truly,' she groaned, as she strove to rise, 'I have been kicked by the Devil himself!'

Now Nightingale's house was built on seven pillars and stretched for seven versts around, and the robber's wife and daughters offered Ilya much golden treasure as ransom for their father. Three thousand gold roubles they offered, but Ilya refused to take them. Thirty thousand gold roubles they offered, but still Ilya rejected them, and not even three hundred thousand gold roubles would he take.

'Take all our red gold, all our pure silver, all our round pearls,' they cried. 'Take as much as you can carry on your trusty steed, only leave us Nightingale, our father.'

But Ilya of Murom spurned their riches, turned his back and led Nightingale away to the famous city of Kiev. A great banquet was in progress when he rode into the wide courtyard, for the Prince and his suite had just returned from the Church of God, and all sat down to eat and drink, and all the noble princes, rich merchants, and powerful bogatyrs were in the palace. Bold Ilya marched into the bright chambers, bowed low on all sides, and was bidden sit as an honoured guest at the oaken table.

'Where do you come from, bold fellow?' asked Prince Vladimir. 'By which route did you come to Kiev?'

'My name is Ilya, son of Ivan,' said the hero. 'I come from the village of Karacharov by Murom, and I have taken the straight road to Kiev from there.'

A roar of protest echoed through the hall.

'Fair Sun, Prince Vladimir!' cried the mighty bogatyrs. 'This man lies! How could he have come by the straight road to Kiev when no horse, no wild beast, and no black raven has ever passed that way for thirty years? Tatars besiege the city of Smolensk, the path is barred by the Black Mire and the raging Smorodina River, bandits lie in wait in the woods of Brianski, and for thirty years the Robber Nightingale has sat on the tops of seven oaks! He pipes like a nightingale, hisses like a serpent, roars like a bull, and shrieks like a Tatar, and all the green grass flattens, the blue flowers scatter, the dark forest bows to the ground, and all men fall dead!'

'Fair Sun, Prince Vladimir,' said Ilya of Murom, 'I do not lie. I have come by the straight road to Kiev, and Nightingale the Robber stands outside in your wide courtyard, tied to my steel stirrup by his golden curls!'

Prince Vladimir rose to his sturdy feet, threw his mantle of marten skin over his shoulder, placed his cap of sable skin on his head, and went out of the hall into the wide courtyard to see the famous Robber Nightingale. And when the Prince desired the robber to sing, Ilya of Murom ordered him to obey the Prince's command.

'But do so softly and at half strength, bold Nightingale,' he warned, 'or you shall answer again to me!'

But the robber paid no heed to Ilya's warning, and opening his mouth, he piped like a nightingale, hissed like a serpent, roared like a bull, and shrieked like a Tatar. The echo roared through the whole town of Kiev, the old towers crumbled, the new towers shook, the church-domes twisted, the windows smashed and scattered over the streets, and mighty boyars, brave bogatyrs, and rich merchants dropped dead on the spot! And had not Ilya covered Vladimir's head in his sable cap and marten cloak, the Prince of Kiev would have surely died!

'Fair Sun, Prince Vladimir!' said Nightingale. 'Ilya of Murom, old Cossack! Let me free, and with my wealth I shall build towns and villages all around the famous city of Kiev, and holy monasteries for the Christian faith.'

'Beware, Prince Vladimir,' said Ilya of Murom. 'The like of him buildeth not, but destroyeth ever!'

'If you let him go, Prince Vladimir,' cried the bogatyrs, 'he will kill us all!'

So Ilya took Nightingale the Robber by his white hands and dug a deep pit one hundred fathoms long and broad, and cutting off his turbulent head, he threw his body into the pit.

'Be our elder brother, Ilya of Murom, old Cossack!' cried the bogatyrs. 'Be the defender of the Russian land!'

And a great banquet was prepared in his honour, and all bowed low before him.

'Live in our famous town of Kiev, old Cossack,' cried Prince Vladimir, 'and we shall reward you well.'

The Three Journeys of Ilya of Murom

ILYA OF MUROM's youth had turned to old age and his old age to the shadow of the tomb, and as he journeyed on across the open plain, he came to a point where three roads met, and there at the cross-roads stood an oaken sign-post inscribed thus in letters of gold:

'He who takes the first road shall be slain; he who takes the second road shall be wed; and he who takes the third road shall be rich.'

'What should it avail an old man like me to be wed?' thought the hero. 'And what use are riches to the aged? I shall take the first road, for Death alone befits the old.'

Ilya beat his horse on its sloping flanks, and when he had galloped along the first road for the space of three hours and a distance of three hundred versts, he arrived at the foot of a high hill upon which a large white palace stood. Now this palace was the abode of robbers, and they were not less than forty thousand in number. When they spied the old Cossack riding below, they scrambled quickly down the mountain-side and tried to drag him from his horse.

'Ho there, thieves and robbers, rogues and highway-men!' called out Ilya. 'What do you want of an old man like me? I have neither gold nor silver nor the smallest shining

43

pearl. All that I have is my trusty steed, and he is beyond all price. Precious jewels are woven in his mane and shining stones in his tail—not for the sake of grace or beauty, but to shine in the dark autumn nights that I may see for fifty versts around. But my Circassian saddle is worth five hundred roubles and three hundred roubles my Christian cross!'

'Ho there, thieves and robbers, rogues and highwaymen!' cried the robber ataman in a terrible voice. 'Why do you waste time parleying with an old man? Drag him down from his horse and strike off his turbulent head!'

'Ye shall not kill or rob me, old Cossack,' cried Ilya, as the robbers rushed upon him, 'though I wear a cloak of marten skin with three hundred buttons on it worth fully eight hundred roubles!'

The robbers were angry at his taunts and rose all around the old Cossack, trying to drag him down.

'Ye shall not kill or rob me, old Cossack,' cried Ilya. 'My trusty steed, which no money can buy, leaps across mountains and bestrides the mighty rivers!'

But still the infuriated robbers strove to kill him.

'Ye shall not kill or rob me, old Cossack,' cried Ilya, 'for I have a trusty bow and still ten arrows left!'

But still the robbers came on and on.

'Who among ye has little children?' cried Ilya. 'Who among ye has a fair young wife? If any there be, let them think now of their widows and orphans, for ye are about to perish all!'

And stretching his mighty bow, the old Cossack shot an arrow into the ground, and so great was the force thereof that the yellow sand flew up in clouds and the damp earth shook.

When they saw this, the robbers stopped in their tracks, and trembling with fear, fell upon their knees.

'Do not slay us, Ilya of Murom!' they cried. 'Let us live, and we shall give you golden treasure, flowered garments and herds of fine horses!'

'If I took your golden treasure,' replied Ilya, 'I should

needs dig vaults to hold it. If I took your flowered garments, I should needs hire carts to bear them away. If I took your herds of horses, I should needs become a herdsman, and I still have a long way to go alone.'

And rising in his saddle, he brandished his keen steel sabre, and where he slashed, a street appeared, and where he struck, a road. And Ilya of Murom cut off the heads of the robbers like thistles, from the first till the last, till none remained, and riding back to the cross-roads, he wrote thus upon the wooden signpost:

> *'Ilya of Murom, old Cossack, rode this way, and was not slain; and thus is the first road cleared for ever.'*

Then Ilya beat his horse on its sloping flanks, and when he had galloped along the second road for the space of three hours and a distance of three hundred versts, he came to a place too large to be a village and too small to be a town, and he halted his horse before a palace of white stone. Now there dwelt therein a fair young maiden, and when she saw the bold Russian knight, she came out of the palace to greet him, and curtsying low before him, she took him by his white hands, kissed him on the lips, and led him into the palace of white stone. She sat him at the oaken table, covered it with a silken cloth, placed upon it all manner of sweet viands, wines, and honeymead, and bade him eat and drink.

And when the old Cossack had eaten and drunk his fill, the maiden took him by his white hands and led him to a soft bed in a warm bedchamber, and bade him rest upon it. But Ilya sensed her perfidy, and lifting her in his strong arms, he dropped her heavily on the soft bed in his place. And straightway the treacherous bed fell open, and the fair maiden, caught in her own trap, fell straight down to the dungeons below. And taking her golden keys, Ilya went down and opened the iron gates of the dungeons and let out the multitude of kings and princes and mighty Russian bogatyrs who had fallen into the treacherous maiden's hands.

'We thank you, Ilya of Murom,' they cried, 'for you have saved us from our doom.'

And leaving the fair maiden in the deep dungeon alone, Ilya rode back to the cross-roads whence he came, and wrote thus upon the wooden signpost:

> *'Ilya of Murom, old Cossack, rode this way, and was not wed; and thus is the second road cleared for ever.'*

Then Ilya beat his horse on its sloping flanks, and when he had galloped along the third road for the space of three hours and a distance of three hundred versts, he came to a huge rock in the middle of a field. Now the rock weighed thrice times ninety pood, but the old Cossack dismounted from his horse and pushed the stone with his mighty shoulder. The great stone rolled aside, and there in the pit lay gold and silver, precious gems and large round pearls. And taking the treasure, Ilya shared it out among the poor and the widowed and orphaned, and building also a church with many bells, he rode back to the cross-roads whence he came, and wrote thus upon the wooden signpost:

> *'Ilya of Murom, old Cossack, rode this way, and was not made rich; and thus is the third road cleared for ever.'*

Stavr Godinovich and his Clever Wife

A<small>T</small> the palace of the gentle Prince Vladimir in the capital city of Kiev a great banquet was in progress and all the princes, boyars, and bogatyrs were gathered together. As the day turned to evening and all had eaten and drunk their fill, the merriment of the feast increased and the guests began to make their boasts. Some boasted of their fame or their sturdy youth, others of their riches or their trusty steeds, but the brave warrior Stavr Godinovich sat at the oaken table and boasted of nothing.

'Well now, Stavr Godinovich,' cried Prince Vladimir, perceiving his silence. 'Why do you sit so woebegone and sad at my merry feast? Why do you hang your head? Have they given you a place unworthy of your rank, or have they passed you the cup out of turn?'

'Great Prince,' replied Stavr Godinovich, 'I am well content with my place at your table, and am in no way offended. But alas! I have nothing to boast about. I have no family and no father or mother, for I am an orphan. I have no great riches and have no trusty steed. I have only a fair young wife, who is as brave and skilful as an Amazon. How well does she wield the bow, for when she fits the arrow to the silken string and shoots, she never misses! Place a steel knife behind a golden ring and she will shoot her arrow through the ring and split it in two against the knife. And so enchantingly does my young wife play on the maplewood psaltery, and so cunningly does she play at chess, that all good people are amazed. In one hour she could confound all the Russian knights, and in one day she would defeat the great Sun, Prince Vladimir, himself!'

This talk did not please the Prince of Kiev, and his face grew pale with anger.

'Ho there, my trusty servants!' he cried. 'Seize this youth by his white hands, throw him into the deep dungeons, and give him but oats and water for food, that he may boast no more such idle boasts!'

The soldiers ran up and dragged Stavr Godinovich off to the deep dungeons, shut him behind great oaken doors, and closed them with great iron locks, and Stavr sat in the darkness engrossed in melancholy thoughts, and sorrowed greatly that he had offended the Prince with a careless boast. As he looked out of the tiny window of his cell, however, the fair sun shone in and melted his fears, and he began to sing a happy song. Just at this moment there chanced to pass by his window the beautiful maiden Zabava Putyatichna, the beloved niece of Prince Vladimir, and hearing the sound of singing, she approached the tiny window of the dungeon.

Greetings, brave and goodly youth,' she said, thinking him to be a pagan prisoner. 'From what country do you come, from what horde? Who are your father and mother?'

'Alas! fair maiden,' replied Stavr Godinovich, 'I serve the great Prince Vladimir, and have done so truly and faithfully for nine whole years. And my wages are this dark dungeon, and oats and water my food.'

'What is your name, brave youth?' asked Zabava Putyatichna.

'Stavr Godinovich is my name,' he replied.

Hereupon the maiden left him, but when the golden sun had set and the white light of day turned to shadows, Zabava crept up to the narrow window and lowered food down on a rope that Stavr Godinovich might eat his fill.

In the meantime, Stavr's young wife, Katrina Ivanovna, sat alone in her marble chamber in her palace of white stone and ate and drank and was happy, for she knew nothing of her husband's sorry plight. As she opened her casement window, however, a large bird flew into the room, a black raven, and perched itself upon the sill.

'Why do you sit thus, young wife?' asked the bird in a human voice. 'You eat and drink and make merry as though no misfortune had fallen upon you, and yet your husband, brave Stavr Godinovich, has been thrown in the dungeons of the palace of gentle Prince Vladimir of Kiev.'

Katrina felt neither fear nor despair, but leaping to her sturdy legs, she said:

'Thank you, black raven, wise bird. I have now no time to stop and speak with you. If you have spoken true, you shall be my guest; but if you have spoken false, I shall cut off your head!'

And going down into the wide courtyard, she summoned her brave druzhina of thirty archers, thirty chess-masters, and thirty minstrels, and addressed them thus:

'Prepare yourselves for the road, my trusty druzhina, true and faithful servants, for we are going to Kiev, to the palace of gentle Prince Vladimir, to rescue my husband, Stavr Godinovich, from the dungeons.'

And going to her stables, she led out her black steed, and harnessed it for the road. She placed upon its back the fine Circassian saddle with the silken girths and golden stirrups—not for the sake of beauty, but for strength—and tying the horse to the golden ring in the wooden post in the courtyard, she returned to her chamber, dressed herself in her husband's clothes, took leave of her old mother, and donned the armour of the Russian bogatyr. She took the strong bow and the well-tempered arrows; she took the club of steel and the long sharp lance. She carried a falcon on her left hand and a dove on her right hand, and mounting her trusty steed, she rode out into the open plain to join her druzhina, and all set out for Kiev.

When Katrina Ivanovna and her men came within three

furlongs of the walls of Kiev, she bade her druzhina rest
from the long journey, and riding herself into the city, she
announced her presence to the great Prince Vladimir.
Katrina walked into the great banqueting hall, and the rafters
shook and the floorboards trembled, and all marvelled to see
such a mighty warrior, for they took her by her manner and
dress to be a man. She crossed herself, bowed low on all
sides and to Prince Vladimir, and addressed the great prince
thus:

'Prince of Kiev,' she said, 'I am the Ambassador from
the land of Greece, from its King, the dog Kalina. I have
been sent to you to exact tribute for my King, and if you
refuse to pay, we will wage a wicked war against you, for
not far off in the open plain stands my army of forty thousand
men!'

The sturdy legs of Prince Vladimir shook, his bright eyes
clouded over, his sweet lips trembled, and he could not find
words to answer.

'Prince Vladimir,' said the Ambassador, 'I can wait no
longer. Give me your answer!'

'Give me time, terrible Ambassador,' pleaded the Prince.
'Grant me three days and three nights to reflect.'

'We have travelled far and are weary,' answered the Ambassador. 'Time is precious. If you cannot pay the tribute,
give me your beloved niece Zabava Putyatichna to be my
wife!'

'My niece Zabava is of marriageable age,' replied Prince
Vladimir, 'but I cannot give her to be your bride without her
consent. Grant me at least one day and one night to reflect!'

'So be it!' said the Ambassador, and promising to return
on the morrow, she bowed, took her leave of the Prince and
returned to her brave druzhina in the open plain.

Prince Vladimir went sadly to his beloved niece Zabava
with the Princess Apraksia and found her sitting at her
window gazing across the open plain at the retreating Ambassador. She watched him as he walked happily back to his
camp, throwing his long spear up to the clouds and catching
it as it fell.

'Beloved niece Zabava,' said Prince Vladimir, 'save us from our dilemma. A terrible Ambassador has come from the King of Greece, the dog Kalina, and exacts much tribute from our land. If I cannot pay it quickly—and I fear I cannot pay it at all—the Ambassador has resolved to take you to be his wife, and we shall never feast or make merry again!'

'Dear uncle,' said Zabava Putyatichna, 'I am obedient to you in all things, and I do not refuse to take a husband. But I beg you not to make yourself the laughing-stock of all Kiev and all Chernigov too. Do not give me a woman for my husband, for your terrible Ambassador of Greece is not a man but a woman!'

'Zabava Putyatichna,' said Prince Vladimir, 'look out of your window and watch the Greek Ambassador as he returns to his army of forty thousand men. He shoots arrows from his stout bow, tosses his spear high in the air, and catches the swift falcon in its flight. Is *that* a woman?'

'Dear uncle, Prince Vladimir!' replied Zabava. 'Can you not see that the Ambassador purrs and lisps with a woman's voice, sidles along with a woman's mincing gait, and has the white and dainty fingers of a woman that still bear the marks of golden rings? Are these signs not enough? The Ambassador of Greece is a woman!'

'You may be right,' said the Prince. 'But what am I to do to see if he be male or female?'

'The Ambassador is a woman and full of woman's wiles,' replied Zabava. 'Ask the Ambassador for yet another day to reflect and then test his skill at all the manly arts. Then you shall soon see whether he be man or woman.'

But when Prince Vladimir asked the Ambassador for more time, the latter grew very angry.

'Tell me here and now whether we shall receive the tribute money,' she cried, 'for that is the reason we came to Kiev. If you cannot pay it, give me your fair niece Zabava to be my bride!'

'Do not be angry, Ambassador,' said Vladimir, who had no hope of paying the tribute and no wish to give his niece to a Greek. 'Sit at the oaken table with me and we will play the maplewood gusly together. '

'I have my minstrels with me in my white tents,' said Katrina Ivanovna, 'but they are weary of the journey and rest in the open plain. In my childhood, however, I played the gusly myself.'

And taking her maplewood gusly, Katrina plucked softly at the strings and began to sing. And she sang a song of Kiev and a song of Jerusalem, and Prince Vladimir was amazed at her skill.

'Zabava Putyatichna,' said the Prince to his niece, 'not only is the Ambassador a man, but we have none better in the whole of Holy Russia. My minstrels are the best in the land, but none can play the gusly as well as he.'

'The Ambassador *is* a woman,' cried Zabava. 'Do not marry me to a woman! Find out more about the terrible Ambassador of Greece.'

'Welcome guest, terrible Ambassador,' said Prince Vladimir, 'let us play a game of chess.'

'I have my chess-masters in my white tent,' replied Katrina, 'but they are weary from the long journey. In my childhood, however, I used to play chess myself.'

And Vladimir and the Ambassador sat at the oaken table to play a game of chess. The Prince made one move, and was checked. He made another, and was checked again, and Katrina easily outplayed the great Prince Vladimir.

'Zabava Putyatichna,' said Vladimir, 'marry this man. I can no longer put the Ambassador to the test.'

'Take the guest into the open plain, uncle,' said Zabava, now in despair, 'and invite him to show his skill at archery. Then you shall surely see that he is a woman.'

'Let us go into the open fields and try our skill at archery, Ambassador,' said Vladimir.

'I have my archers,' replied Katrina Ivanovna, 'but they are weary from their long journey. But I myself used in my childhood to wield the bow.'

And going into the open fields, they set up a golden ring in the ground and placed a steel knife behind it. Then

Vladimir shot one arrow and missed; he shot a second arrow and missed again; and he shot a third time, but still he could not hit the target. Then Katrina bent her stout bow, drew taut the silken string, and shot her well-tempered arrow at the ring. The arrow hissed like a snake, and flying through the golden ring it cut itself in two equal halves against the edge of the knife.

'Well, Fair Sun Vladimir, Prince of Kiev,' said the Ambassador. 'Are we longer to waste our time thus? Is it not time to settle our business? Pay me the tribute you owe my King, the dog Kalina, and we will depart in peace, or give me your fair niece Zabava to be my bride.'

'Alas, terrible Ambassador,' replied Vladimir, 'I cannot pay the tribute.'

'Then give me Zabava Putyatichna to be my bride,' said the Ambassador, 'for we can wait no longer.'

Prince Vladimir went sadly to his beloved niece.

'Alas, Zabava,' he said. 'If you will not marry willingly, I must marry you against your will.'

'If I live, I shall not marry a woman!' cried Zabava. 'We should be the laughing-stock of all Russia. The Ambassador plays the gusly, plays chess and wields the bow like any man, but he talks, walks, and sits like what he is, a woman!'

'Ho there, servants!' cried Vladimir in a loud voice. 'Dress my niece in flowery dresses, put a golden crown upon her head and golden rings upon her fingers, and take her to the cathedral, for she shall marry the terrible Ambassador of Greece! The whole town of Kiev shall not suffer for the sake of one maiden!'

And to the Ambassador, he said:

'Your bride is prepared and waits for you at the church.'

'Wait, Fair Sun Prince Vladimir of Kiev,' said Katrina Ivanovna. 'Before they place the marriage crowns upon our heads, let us ride out together and pit our manly strength one against the other.'

'Dear guest, terrible Ambassador!' cried Prince Vladimir. 'Neither I nor any of my bogatyrs is any match for you!'

'Are there none in your dungeons you could spare?' asked the Ambassador, and Vladimir remembered Stavr Godinovich.

'If I release Stavr Godinovich,' he thought, 'I need not see him, but if I do not, I shall anger the Ambassador.'

And aloud he cried:

'Ho there, servants! Take these golden keys and open up the deep dungeons, and let out all the strong young men!'

His servants obeyed the Prince's command and led Stavr out into the wide courtyard, took off his convict's garb, dressed him in knight's armour, set him on a good horse, and sent him to do battle with the terrible Ambassador of Greece. And galloping together into the open plain until the dust hid their horses from view, Katrina and Stavr sported together, leapt from horse to horse, threw their steel clubs into the air and caught them as they fell. Then, returning with Stavr Godinovich, the terrible 'Ambassador of Greece' bowed low before the great Prince Vladimir, and casting off her knight's armour, revealed her woman's dress beneath and let down her long hair.

'Great Prince Vladimir, Fair Sun,' laughed Katrina Ivanovna, 'bring my "bride", Zabava Putyatichna, back from the cathedral and return her to the women's quarters, for I would not make a mockery of the girl. Never yet has it been seen or heard that a girl was married off to a woman as you, great Prince, would have done. I have rescued my beloved husband from your deep dungeons, Prince Vladimir, and now, farewell!'

And laughing and sporting together, Stavr Godinovich and his clever wife Katrina Ivanovna rode away, and Prince Vladimir saw to his sorrow and confusion that his trusty knight had made no empty boast.

Sadko the Minstrel

IN the glorious town of Novgorod there once lived a
minstrel named Sadko. He possessed but little gold,
and to earn money he went round the noble feasts and
banquets and enchanted everyone, princes, boyars, and
merchants alike, with his marvellous skill on the gusly, the
Russian psaltery. One day, however, misfortune befell him,
for no one that day summoned him to play at any feast or
banquet. Then a second and a third day passed, and still the
minstrel Sadko remained without hire, and sighing deeply,
he went down to the shores of Lake Ilmen, sat in the shade
of a rock and began to pluck the strings of his maplewood
gusly. He played thus all day, and just as evening was falling,
the waves of the lake suddenly swirled and roared and the
waters were clouded with sand. Sadko was terrified and running away from the lake he returned to the town of Novgorod.

The dark night passed and the light of day came again,
but still no one invited Sadko to play at a feast, and finding
his enforced idleness most tedious, the minstrel went again
down to the shores of Lake Ilmen, sat in the shade of the
burning rock, and began to play. He played all day, and as
dusk fell, the waves of the lake suddenly swirled and roared
and the waters were clouded with sand, and Sadko ran away
and returned to Novgorod. And still on the third day no

one invited Sadko to a feast, and again he went down to the shores of Lake Ilmen and played all day in the shade of the burning rock. And as dusk fell on the third evening, the waves of the lake suddenly swirled and roared and the waters were clouded with sand. This time, however, Sadko was not afraid, and he continued to pluck the strings of the psaltery on the shore of the lake. And then, when he had played thus for some time, the waves of the lake suddenly swirled and roared louder than ever, and before the terrified minstrel could move a limb, the waters parted, and out of the lake stepped the mighty figure of the great King of the Blue Seas!

'We thank you, Sadko of Novgorod,' roared the King, 'for you have greatly diverted us. I have held a great banquet in this Lake of Ilmen, and you have enchanted the ears of my guests with your playing, and all are grateful to you. I know not, Sadko, how rightly to reward you for your services, but go now, return to Novgorod, and tomorrow you shall be summoned to play at a banquet given by a rich merchant. All the rich men of Novgorod will be present, and when they have eaten and drunk their fill, they will all begin to praise themselves and boast. One will boast of his immense wealth, one of his excellent steed, another of his strength and valour, and another of his youth. The wise will boast of his aged father and mother, and the fool will boast of his fair young wife. And you also, Sadko, shall make your boast. In the presence of all the rich merchants you must say: "I, Sadko the Minstrel, know that in Lake Ilmen there are fish with fins of pure gold!" The rich merchants of Novgorod will laugh and contradict you and say that there are no such fish in Lake Ilmen, whereupon you must make a wager with them. Wager your turbulent head against their shops and precious wares, and when they accept, take a silken net and come and cast it into the lake three times, and each time I shall send you a fish with golden fins. In this way you will win row upon row of shops in the market and you, Sadko the Minstrel, will become one of the richest merchants of Novgorod.'

Sadko returned to Novgorod, and lo and behold! on the next day he was invited to a great banquet given by a rich merchant. Everything happened as the King of the Blue Seas had predicted. When all the rich merchants had eaten and drunk their fill, they began to boast of their achievements. The one boasted of his countless treasure, the other of his strength and valour, the wise man boasted of his aged father and mother, and the fool of his fair young wife, and all the while Sadko sat and said nothing. When the rich merchants had finished their boasts, they turned to the silent minstrel and asked him whether he had nothing to boast about, and laying aside his gusly, Sadko stood up and said:

'Alas, merchants of Novgorod, what should the poor Sadko have that he could match his boasts against yours? I have no countless golden treasure, and I have no fair young wife. I have only one thing whereof to boast, for I alone know that in Lake Ilmen there are fish with fins of gold!'

At this the merchants of Novgorod burst out laughing and argued with him, saying there was no such fish in the lake.

'If I were rich,' cried the minstrel, 'I should wager a great sum on the truth of what I say. As it is, I have nothing but my own turbulent head to stake.'

'We shall accept your wager, Sadko!' laughed the merchants. 'We say there are no fish with golden fins in Lake Ilmen, and we wager all our shops in the market and all their precious wares against your turbulent head!'

Hereupon Sadko took a silken net, and going down to the shores of Lake Ilmen, cast it into the waters. He cast it once, and straightway drew out a tiny fish with golden fins. A second time and a third time he cast in his net, and each time he drew out a tiny fish with golden fins. The merchants of Novgorod were amazed, but seeing there was nothing to be done and that the minstrel had spoken true, they gave him their shops in the market with all their precious wares, and Sadko the Minstrel became one of the richest merchants in the glorious town of Novgorod, and began to trade. And he journeyed from place to place and to many towns of Russia near and far, and when he had made great profits, he took a fair young wife, and built himself a palace of white stone and as the red sun, the bright moon, and the silver stars shone in the heavens, so did a sun, a moon, and a host of silver stars shine in the halls of Sadko's palace, for he adorned it with all things of beauty. Then Sadko organized a great banquet and invited to his palace all the rulers and merchants of Novgorod to take part. The governors Luka Zinoviev and Foma Nazariev were there with all the freemen of the town, and sitting down to the feast, they ate and drank their fill, and began to boast their boasts. Then Sadko rose to his feet, and walking through the magnificent halls of his white palace, he addressed his guests thus:

'Powerful lords, rich merchants, and freemen of this glorious town of Novgorod! You are all welcome guests at my banquet. You have eaten and drunk and grown merry, and you have filled your bellies and boasted your boasts. What should I, Sadko, now boast of? My golden treasure is inexhaustible, my flowered robes will never wear out, and my trusty druzhina is incorruptible. I will therefore boast a mighty boast! I wager that my wealth is so great that I can buy up all the wares, good and bad, for sale in the glorious town of Novgorod!'

Then Foma Nazariev and Luka Zinoviev stood up on their sturdy legs.

'Sadko, rich merchant of Novgorod,' they said, 'how much do you wager?'

'Rulers of Novgorod,' replied Sadko, 'what would you that I wager of all my countless treasure?'

'Let your wager be thirty thousand roubles.'

So Sadko the Merchant wagered thirty thousand roubles and the banquet came to an end.

When Sadko awoke the following morning, he roused his trusty druzhina, gave them as much gold as they could carry, and going down to the market, he bought up all the goods of Novgorod, both good and bad. A second and a third day he did the same until not a single thing of Novgorod remained for sale. As soon as the wares of Novgorod itself were exhausted, however, the fine wares of Moscow arrived to fill the market, and Sadko began to reflect on his folly.

'How could I boast to buy up all the wares for sale in the glorious town of Novgorod?' he said. 'Even if I buy up all the wares that now arrive from Moscow, more will arrive from beyond the seas, and how should I, Sadko, buy up all the wares of the wide, white world? Glorious Novgorod is richer than I, and it is better to pay the thirty thousand roubles I wagered than to fritter away all my riches in a fruitless task.'

Thus did Sadko lose his great wager against the wealth of Novgorod, and surrendering the thirty thousand roubles, he built thirty black boats, loaded them with all the wares of Novgorod and sailed away to sell them in the wide, white world. He sailed from Novgorod to the River Volkhov, thence to Lake Ladoga, from Lake Ladoga to the River Neva, and from the River Neva into the deep, blue sea. And sailing to the lands of the Golden Horde, he sold there the wares of Novgorod, and filling his casks with red gold, his barrels with pure silver, and many coffers with fair round pearls, he sailed back across the sea in the direction of Novgorod. But when the ship was in the middle of the ocean, the winds howled, the waves beat, and the sails tore from the masts until the black ships could move no more, and Sadko summoned his faithful crew.

'My good and faithful crew,' he cried. 'We have journeyed far across the sea, but never yet have we paid tribute to the great King of the Blue Seas. And now he threatens us with storms and demands his due. Hurry, take a cask of red gold and cast it into the blue sea!'

And they took a cask of red gold and cast it into the blue sea. But still the winds howled, the waves beat, and the sails tore from the masts, and the ships could not move. Then Sadko ordered a barrel of pure silver to be cast into the blue sea, and after that a coffer of fair round pearls, but still the cold winds howled, the great waves dashed, and the sails tore, and the black ships could make no headway across the face of the ocean.

'Alas, my brave crew!' cried Sadko, 'it is clear that all the treasure we possess will not suffice as tribute to the great King of the Blue Seas. He demands a human life and that alone will pacify him. Let each one of you make a counter of lead, and writing his name upon it, cast it into the blue sea. I, your captain, will make mine of red gold, and whose-soever counter sinks to the bottom of the ocean shall follow as tribute to the mighty King of the Blue Seas.'

And when they threw their counters on to the surface of the sea, all the lead ones bobbed and floated like ducks on a lake, but Sadko's golden counter sank straight to the bottom.

'These lots are not fair!' cried Sadko. 'Let everyone make counters of red gold, and I, your captain, shall make mine of oak!'

But when they cast the counters into the sea, all the golden ones floated on the surface, while Sadko's oaken counter sank straight to the bottom. And though Sadko commanded the crew to make their counters of oak while he made his of lime, still it was his counter that sank to the bottom of the sea.

'Alas!' sighed Sadko. 'There is nothing to be done! It is me that the King of the Blue Seas desires. Come, my good crew, bring me my merchant's ink-well, my swan's quill, and pen and paper bearing my seal.'

And when his men had done as he commanded, Sadko sat on his folding chair at his oaken table and wrote his will. And sharing his possessions among the Holy Church, his younger brothers, his young wife, and his brave and trusty crew, he began to weep.

'Come, my brave and faithful crew,' he said. 'Place an oaken plank on the blue sea that I may cling to it, for thus shall I find death less terrible.'

And taking his maplewood gusly, he bade farewell to his native town of Novgorod, descended on to the oaken plank on the surface of the waters, and watched the black ships move away; and left alone in the middle of the ocean, he fell asleep.

When Sadko awoke, he found himself on the very bottom of the deep ocean. The red sun shone brightly through the clear water, and there in front of his eyes stood a great palace of white stone, and when he went in, he beheld the mighty King of the Blue Seas seated on his coral throne.

'Welcome, Sadko of Novgorod, rich merchant and minstrel!' cried the King. 'You sailed a goodly time on the face of the ocean without paying tribute to its King! However, now you have come in person to pay me tribute, and since you are a great master on the gusly, I would bid you play me something now.'

Sadko took his maple gusly and struck up a merry tune, and the King of the Blue Seas began to dance. And as he danced the whole ocean trembled, the waves on the surface dashed together, and many black ships sank to the floor of the ocean with much treasure and many good true believers who died praying to Mikola Mozhaiski to save them from the cruel sea. As he was playing, Sadko felt a hand on his right shoulder, and turning his head, he saw behind him a small, white-haired old man.

'You play well, Sadko of Novgorod,' he said. 'These are gay tunes for the Kingdom of the Blue Seas.'

'Alas, batiushka,' said Sadko, 'I would fain play a sad song, for I have no freedom in the blue sea, and it is the King who bids me play for him.'

'If you wish to escape, Sadko,' replied the old man, 'I can help you.'

And thereupon the old man whispered in the minstrel's ready ear what he had to do to escape from the watery domain in which he now found himself.

'And when you return to Holy Russia,' concluded the old man, 'build there a cathedral in the name of Mikola Mozhaiski, for I am he.'

And having said this, the old man disappeared and Sadko was left alone. Thus, remembering what Mikola had told him to do, Sadko broke both the strings and pegs of his

psaltery, the music ceased, and the mighty King of the Blue
Seas could dance no more.

'Why do you play no more, Sadko?' asked the King.

'The strings of my gusly have snapped, Your Majesty,'
replied the minstrel, 'and the pegs have come apart.'

'Ah well,' sighed the King. 'I suppose it cannot be helped.
You have already afforded me much pleasure, and I should
like to make you happy in the Kingdom of the Blue Seas.
Would you not like to marry one of our beautiful sea-
maidens and settle for eternity here?'

'I bow to your will, Your Majesty,' replied Sadko, for
Mikola Mozhaiski had told him it would be perilous to
refuse.

'Very well,' said the King. 'Tomorrow you shall choose
for yourself a bride who shall be your equal in wit and
reason.'

All came to pass as Mikola Mozhaiski had said. A great
parade of all the eligible sea-maidens of the Kingdom of the
Blue Seas was held on the following day, and although
Sadko was much struck by their beauty, he kept his passions
in check, and suffering the first three hundred maidens to
pass by, and then the second three hundred, he finally
chose the last maiden of the third three hundred, a dark and
beautiful maiden named Chernava, just as Mikola had told
him to do. The King of the Blue Seas then gave a great
banquet to celebrate the marriage, and when the time came
for Sadko to retire with his new wife, he remembered the
old man's warning that if he touched her at all, he would
never see his motherland again. So, forbearing to kiss or
embrace his bride, Sadko dropped off into a deep sleep, and
when he awoke, he found himself back on dry land on the
outskirts of Novgorod, and there, sailing along the River
Volkhov, came his own black ships with his good and faithful crew. Joyously Sadko ran to meet them, and when they
saw him waving from the high banks of the River Volkhov,
his men were amazed, for they knew that they had left their
master to die in the middle of the blue ocean, and here he
was in Novgorod before them! They rejoiced greatly to see

him and embraced him, and all went together to Sadko's white-stone palace where he kissed his young wife and related all that had happened to him. Then Sadko unloaded his ships of their countless golden treasure, and with it he built one great cathedral church in the name of Mikola Mozhaiski and another to the glory of the Blessed Mother of God, and praying to God to forgive his sins, Sadko sailed no more out into the blue sea, but spent the rest of his life peacefully in his native town of Novgorod, beloved and praised by all.

Vasili Buslayevich

IN the great and glorious city of Novgorod old Buslay
lived for ninety years in peace and never once had he
quarrel with the men of Novgorod. But Buslay grew
old and died, and left behind him a large estate, a widow
Amelfa Timofeyevna and a young son Vasili Buslayevich
who had not the peaceable nature of his father. When the
lad reached the age of seven, his mother, the widow Amelfa
Timofeyevna, sent him to learn to read and write, and he
learned these things to perfection. They gave him a pen,
and he wrote a skilful hand; they taught him hymns, and he
sang them beautifully, and never was there such a singer in
the town of Novgorod as Vasili Buslayevich. When he grew
up, however, the lad began to consort with drunken and
foolish fellows, who were also brave and merry fellows, and
he would drink until he was drunk and would then begin to
roam about the streets. If he grasped men by their hand,
their arm would be rent from their shoulder; if he seized men
by their foot, their leg would be torn from the rump; and
if he hugged men round the body, they would scream and
roar and break in two. When the townsfolk of Novgorod
could bear this no longer, they banded together, went to the
widow Amelfa Timofeyevna and complained of her son's
behaviour. The widow scolded and upbraided the youth and
tried to bring him to his senses, but Vasili resented the rebuke,

went up to his chamber, sat on his leather chair, and speedily composed a number of well-written letters which read as follows:

'Whosoever would partake of ready food and drink, let him come to Vasili in the wide courtyard of his mansion, and then he may eat and drink and wear flowered clothes at my expense.'

And sending his servant with these letters through the wide streets and narrow lanes of Novgorod, he placed a large barrel in the middle of his courtyard, filled it with green wine, and stood a cup that would contain one and half bucketsful beside it.

Now those men in Novgorod who could read read Vasili's letters aloud to those who could not, and before long a great crowd had assembled in the courtyard. Then Vasili announced that whosoever could raise the cup in one hand and quaff it without pausing for breath should stand the next test, and whoever passed both should become a member of his druzhina and eat and drink to his heart's content.

The first to take the mighty cup in hand and quaff it without pausing for breath was Kostia Novotorzhenii, and when he had done this, Vasili Buslayevich took his great red elm half-filled with lead and weighing twenty pood in his mighty hand, brandished it in the air, and brought it down with all his might on Kostia's turbulent head. But Kostia stood firm and did not waver, and the heavy blow did not so much as ruffle the curls on his head.

'Welcome, Kostia Novotorzhenii!' cried Vasili Buslayevich. 'You shall be called my brother and shall be much more than a blood-brother to me.'

And after Kostia two merry boyar brothers, Luka and Moisei, stood the test, and Vasili was pleased to welcome them as members of his druzhina. Then came the men of Zalyoshen, and after them the seven Sbrodovich brothers, and Vasili could not harm them with his mighty red elm. And so they came on until there were thirty of them save one, and the thirtieth was Vasili Buslayevich. And whoever else tried the test was killed outright and his body thrown over the wall.

When Vasili had chosen his trusty druzhina in this manner, he chanced to hear that the citizens of Novgorod had organized a merry evening with much barley beer, and he went with his men to Nikolay's fraternity.

'We are prepared to pay no little contribution to partake of the feast,' said Vasili, 'for I shall pay you five roubles for each of my brothers.'

And Vasili paid five roubles for each member of his druzhina and twenty-five roubles for himself, and they were admitted to the fraternity. They began to drink the good barley beer and green wine, but when evening came, young and old began to quarrel, and a fight broke out in one corner. Vasili Buslayevich tried to separate the fighters, but one knocked him off his feet and another wrapped his arms round his throat.

'Ho there, Kostia Novotorzhenii!' cried Vasili. 'Luka! Moisei! I am stifling!'

And quickly the two brothers cleared a path through the crowd and struck many dead, wounding twofold and three-fold and breaking arms and legs until the townsmen screamed and begged for mercy.

'Ho there, men of Novgorod!' cried Vasili Buslayevich. 'Since you wish to fight with us, I will lay a large wager. I and my brave druzhina hereby challenge the whole of Novgorod to a fight! If you beat us, I shall pay you tribute to the day of my death, three thousand roubles every year; but if we beat you, Novgorod shall pay me and my men a like tribute to the day of our death!'

The men of Novgorod accepted the challenge, and on the next day the fight began. All the men of Novgorod, all the rich merchants, rushed upon young Vasili Buslayevich and fought from morn till eve, and Vasili and his trusty druzhina struck many dead in Novgorod. Seeing that they were losing the battle, the men of Novgorod conferred among themselves, and bearing great gifts, they went to Vasili's mother, the widow Amelfa Timofeyevna.

'Good widow Amelfa Timofeyevna!' they cried. 'Accept our precious gifts and keep your son from us.'

Taking the gifts, Amelfa Timofeyevna sent the black-haired handmaiden to her son, and she seized Vasili by his white hands and dragged him back to his mother, who gave him a drug to drink. When he was asleep, the old and foolish woman shut him in a deep dungeon, and closed the iron doors and locked them with steel locks.

All this while Vasili's brave druzhina fought with the men of Novgorod, and when the black-haired handmaiden walked along the Volkhov River, the young heroes pleaded with her.

'Help us, black-haired maid!' they said. 'Do not betray us in our need, for death is upon us!'

Then the black-haired maiden cast aside her maplewood buckets, and taking her yoke of cypress wood from her shoulders, she brandished it among the men of Novgorod until the dead lay thick around her. Then, when she began to pant for breath, she ran off to the dungeon where Vasili Buslayevich lay, struck off the steel locks, and tore open the iron door.

'This is no time for sleeping, Vasili!' she cried. 'Your brave druzhina has wounded and slain many men of Novgorod, and many are the turbulent heads they have broken with their clubs. But make haste, for they are weakening fast and will lose the fight!'

And waking from his drugged sleep, Vasili leapt into the wide courtyard, and finding his iron mace not to hand, he seized the axle of an ox-cart and rushed swiftly through the streets of Novgorod. But as he ran to the aid of his men, an ancient pilgrim stood in his way and barred his path. And on his mighty shoulders covering his head he carried a huge bell that weighed fully three hundred pood.

'Stop, young Vasili!' cried the old man. 'Do not flutter so your wings, young fledgling, for you shall no longer drink of the Volkhov river or slay the men of Novgorod. Many fine heroes are against you, and we shall stand firm!'

'Beware, ancient pilgrim!' cried Vasili. 'I have made a

wager with all the men of Novgorod excepting only those of the holy monastery, excepting you too, ancient pilgrim. But since you have provoked me, you shall die!'

And lifting the mighty axle-tree above his head, he smote the bell a fearful blow; but though the ancient pilgrim shook, he did not fall, and when Vasili looked beneath the bell, he saw no eyes in the old man's head!

Vasili hurried on down towards the Volkhov river, and joining the young heroes of his brave druzhina, he fought furiously against the men of Novgorod from morn till eve. Then, seeing they were losing the battle, the men of Novgorod signed papers, and filling a cup with pure silver and another with red gold, they went to the widow Amelfa Timofeyevna.

'Good widow, matushka!' they said. 'Accept our precious gifts and keep your dear son Vasili Buslayevich from us. We will gladly pay three thousand roubles a year. Every year the bakers of black bread shall bring you a black loaf, the bakers of white bread a white loaf, and all the tradesmen —save only the priests and deacons—shall give you of their wares.'

Then the widow Amelfa Timofeyevna sent the black-haired handmaiden to bring back Vasili Buslayevich and his brave druzhina, and running panting through the dense throng of men of Novgorod on the street, she took Vasili by his white hands and told him of Novgorod's surrender. Then she led the victorious Vasili and his men back to the wide courtyard and brought them all green wine to drink, and sitting round in a circle, they sat and drank. And Vasili Buslayevich ordered the men of Novgorod to be entertained in his mother's house, and they brought him presents, one hundred thousand at once. And thus was peace made between Vasili Buslayevich and the men of Novgorod.

Vasili Buslayevich's Pilgrimage

ON the famous lake of Ilmen by the glorious town of
Novgorod swam a grey drake, diving like an angry
mallard, and there too sailed the dark-red ship of
Vasili Buslayevich with his fearless druzhina on board, thirty
valiant men in all. Kostia Nikitin guarded the stern, Tiny
Potania stood on the prow, and Vasili paced up and down
throughout the ship.

'My brave and shining bodyguard, you thirty valiant
youths!' he cried. 'Steer the ship across Lake Ilmen, my
lads, and make for Novgorod.'

The dark-red ship sailed to the shore of the lake and
anchored at Novgorod, and throwing gang-planks on to the
steep bank, Vasili and all his brave men, save only the guards
they left behind, marched down towards his noble palace.
Here Vasili Buslayevich embraced his noble mother, the
widow Amelfa Timofeyevna, and asked for her blessing.

'Give me your blessing, Mother,' he said, 'for I and my
good druzhina are going to Jerusalem to pray to the Lord,
to bow down to the Holy of Holies, and to bathe in the River
Jordan.'

'My sweet wonder, Vasili Buslayevich!' replied his
widowed mother. 'If you go to perform good deeds, I shall

give you my blessing, but if you go for deeds of plunder, I shall not give you my blessing, but shall pray that the earth cease to bear my son!'

But stone burns in fire, steel melts in flames, and a mother's heart relents. Amelfa Timofeyevna gave her son lead and gunpowder, weapons and stores of bread, but bade him restrain his turbulent head. The young men made their preparations, and bidding the widow farewell, they boarded their dark-red ship, hoisted the white linen sails, and sailed away across Lake Ilmen. They sailed for one day, and then another; they sailed for one week, and then another, and then there sailed to meet them many merchants in their ships.

'Greetings, Vasili Buslayevich!' they cried. 'Whither are ye bound?'

'Hail there, seafaring merchants!' replied Vasili. 'I journey not for my pleasure. In my youth I killed and robbed many men, and in my old age I must think to save my soul. Tell me, good merchants, what is the direct course to the Holy City of Jerusalem?'

'Vasili Buslayevich,' replied the merchants. 'By the direct route Jerusalem lies but seven weeks from here, and by the indirect route, a year and a half. Alas! in the famous Caspian Sea, on the island of Kuminsk, three thousand Kazak atamans bar the way, robbing all vessels that pass by and pillaging the dark-red ships.'

'Bah!' cried Vasili Buslayevich. 'I do not believe in dreams or superstitions, but only in this good red elm which is my club. Farewell, merchants! Set sail, brothers, for Jerusalem, and take the direct route!'

When they had sailed a little way, Vasili spied a high mountain, and anchoring by its steep shores, he disembarked, and began to climb it with his trusty bodyguard at his heels. Half-way up the mountain, which was called Sorochinski or Saracen Hill, he came across a heap of human bones and an empty skull lying in his path, and Vasili drew back his foot and kicked them out of the way.

'Why do you thrust me aside and despise me, Vasili Buslayevich?' said the skull. 'I was once no worse a man than you. Beware, Vasili, for on Saracen Hill, where my skull now lies, shall lie the head of Vasili Buslayevich!'

'Either the Devil speaks in you, Skull,' cried Vasili, 'or an unclean spirit!'

And he spat on the bones and went his way.

When they had climbed to the top of the tall mountain, they found a huge stone three fathoms high, three arshins and a quarter wide and a whole axe-throw across, and the stone bore these words upon it:

> '*Whosoever shall divert himself upon this stone and dive headlong across it shall break his turbulent head in twain.*'

Vasili paid but little heed to the warning, and he and his brave retinue began to divert themselves upon the stone and laughed and jumped boldly across it, but none of them dared dive headlong across. Then they descended the mountain, returned to the dark-red ship, hoisted the fine linen sails, and made for the Caspian Sea. Soon, however, they arrived at a great boom placed across the sea by the Kazak robbers, and there on the harbour stood one hundred of them. But young Vasili sailed boldly into the harbour, cast the gang-planks out on to the steep shore, and leaping on to the dry land, there he stood, leaning calmly on his great red elm. At this the Kazak guards, though they were brave fellows, were greatly dismayed, and without waiting for him to approach, they fled quickly to their atamans and told them of the fearsome Russian who had arrived on the quay.

'For thirty years we have held this island,' said the atamans, 'and never once have we beheld such terror in our men. Who can this Russian be but Vasili Buslayevich, for he flies like a falcon and prances like a hero?'

Vasili approached the Kazak atamans, and they came towards him and stood round him in a circle.

'Greetings, Kazak atamans,' he said, bowing to them. 'Tell me the straight road to the Holy City of Jerusalem.'

'Hail, Vasili Buslayevich,' replied the robber chieftains.

'Pray sit first at our table, be our guest and eat our bread.'

Vasili did not refuse, but sat down with them at one table. They poured him out a goblet of green wine, one and a half bucketsful, and Vasili took the cup and quaffed it at one draught. The atamans marvelled greatly at this, for they themselves could not drink more than half a bucketful, and Vasili and the robbers ate bread and salt together. Then the Kazak atamans gave him presents—the first a bowl of pure silver, the second a bowl of red gold, and the third a bowl of large white pearls. And thanking them for these gifts, Vasili bowed and asked them also for a guide to lead him to Jerusalem. And the atamans dared not refuse his request, but gave him a young guide and bade farewell. Vasili boarded the dark-red ship, the white linen sails were hoisted, and away they sailed across the Caspian Sea until they came to the River Jordan, where dropping their heavy anchors and lowering the gang-planks they stepped ashore in the Holy Land. Vasili Buslayevich entered Jerusalem with his trusty bodyguard, worshipped in the Church of the Holy Sepulchre and ordered masses for his mother's health and his own, and masses in memory of his father and forbears. The next day he purchased masses for the brave young warriors whom he had robbed and slain in his youth, and bathed in the River Jordan. And the priests and the deacons he paid well indeed, and he gave his golden coins unstintingly to the old people at the church.

And while his trusty druzhina were bathing in the Holy River Jordan, an old crone chanced to pass by.

'Why do you bathe in the River Jordan?' she said. 'None may bathe there save Vasili Buslayevich himself, for in the River Jordan our Lord Jesus Christ was baptized. But beware, for you shall lose him, your great ataman, the brave Vasili Buslayevich!'

'Bah!' said the guards. 'Our Vasili believes in none of these things, neither in dreams nor superstitions!'

When his prayers and devotions were done, Vasili returned to his men and ordered them to regain the ship and set sail from the mouth of the River Jordan. They hoisted the white sails of the ship, sailed away to the Caspian Sea, stopped awhile with the Kazak atamans on the island of Kuminsk, and then set sail again for Novgorod. When they had sailed one week and then another, Vasili spied again the tall Saracen Hill, and again they stopped there and climbed up. And when he was half-way up the mountain-side, Vasili came once more upon the heap of human bones and the empty skull, and again he drew back his foot and kicked them aside.

'Why do you again thrust me aside and despise me, Vasili Buslayevich?' said the skull. 'I was once no worse a man than you. Beware, bold Vasili, for on Saracen Hill, where my skull now lies, shall lie the head of Vasili Buslayevich!'

But Vasili spat and went his way, and when they came to the great stone on top of the mountain, Vasili paid no heed to the warning engraved on the stone but began to divert himself with his druzhina, and they leaped and jumped across the stone. But this time Vasili desired to dive across the great stone, and taking a long run, he sped forward and dived headlong across. But as he leapt, his foot stumbled, and when he was barely a quarter of the way across, his head struck the stone, his brains were spilled, and he died. And mourning deeply, his comrades took his body and buried him in the place of the empty skull, and running down from Saracen Hill, they boarded the dark-red ship and sailed home to Novgorod, and going to the widow Amelfa Timofeyevna, they gave her a letter from her dead son. When the widow read the letter, she burst into tears.

'O brave young men!' she said. 'This is my son's will. Go down into the deep vaults beneath this white-stone palace, and take much golden treasure without stint.' And the black-haired handmaiden led Vasili's men down to the vaults, and each taking a little of the golden treasure, they returned to the widow.

'We thank you, Amelfa Timofeyevna,' they said, 'for you have given us food and drink, and shoes and clothes.'

And the widow Amelfa Timofeyevna ordered a goblet of
green wine to be poured out for each one, and the black-
haired maiden brought it to the brave young heroes, and
they drank and bowed and took their leave and went each
his way.

Since when there have been no more Heroes in Holy Russia

WHEN the sun was setting in the West, seven of the bold heroes of Russia—Godenko Bludovich, Vasili Kazimirovich, Vasili Buslayevich, Ivan Gostyny Syn, Alyosha Popovich, Dobrynia Nikitich, and the old Cossack, Ilya of Murom—assembled together and rode out towards the River Safat. The open plain stretched before them, and before long they came to an old, withered oak-tree which marked the point where three roads led in different directions—the first to Novgorod, the second to the famous city of Kiev, and the third to the blue sea. Now the third road had been shut off for thirty years, nay, thirty years and three, and the seven bogatyrs halted at the cross-roads and prepared to rest for the night. The Russian heroes went to sleep in their white linen tents, while the horses, chinking their golden bridles, wandered at large over the silk-green grass.

As soon as the red sun rose in the heavens, bold Dobrynia Nikitich rose early before the others were awake and went down to the River Safat, and when he had washed himself in the ice-cold water, dried himself on the towel of thick linen, and prayed to the holy icon, he suddenly espied a white tent pitched on the other side of the stream. Now in that tent dwelt an evil Tatar, a wicked Musulman, who permitted no Russian, on horseback or on foot, to cross that river. Bold Dobrynia saddled his swift steed, placed upon it the well-woven sweat-cloth and fine Circassian saddle, grasped his lance and sword in hand, and swung himself on to the back of his horse. The horse reared beneath him, and leaping from the damp earth, with one bound covered a whole verst. Riding up to the white tent, Dobrynia called out in a loud voice:

'Come out, you evil Tatar, you wicked Musulman, and do honourable battle with a Russian bogatyr!'

When he heard these words, the Tatar rushed from his white tent, leapt on to his horse, and like two great whirlwinds, like two mighty storm clouds, the two warriors clashed. Their sharp spears splintered and their steel swords smashed, but the two heroes jumped down from their steeds and continued the fight with their fists. And as they fought, Dobrynia's right foot slipped, his right hand faltered, and he fell to the damp earth. The Tatar flung himself upon him, trampled him underfoot, ripped open his white breast, and tore out his heart and liver. Bold Dobrynia, victor of so many frays, was dead!

In the meantime, as the red sun rose higher in the heavens, bold Alyosha Popovich rose while the others still slept, and going down to the River Safat, he washed in the ice-cold water, dried himself on the towel of thick linen, and prayed to the holy icon. And suddenly Dobrynia's horse stood before him, fully saddled and bridled, its sad eyes cast on the ground, for it grieved for its dead master Dobrynia Nikitich. Alyosha Popovich, sensing disaster, saddled his trusty steed and rode forward.

A white speck shone in the plain, the white of the Tatar's tent; a blue flash shone in the plain, the blue of keen steel swords; a red patch shone in the plain, the red of a hero's blood. And as Alyosha rode up to the white tent, he saw brave Dobrynia lying on the ground; his clear eyes stared blindly, his powerful hands were limp, and blood clotted his white breast, and Alyosha grew pale with anger and called out in a loud voice:

'Come out, you cursed Tatar, wicked Musulman, and do honourable battle with a Russian bogatyr!'

Hearing these words, the Tatar rushed from his white tent, leapt on his horse, and like two great whirlwinds, like two mighty storm clouds, the two warriors clashed. Their sharp spears splintered and their steel swords smashed, and the two heroes jumped down from their steeds and continued the fight with their fists. And Alyosha's right foot slipped not, neither did his right hand falter, and he threw the Tatar down on the damp earth, trampled him underfoot, and made to rip open his white breast and tear out his heart and liver. But as his knife was poised in readiness, a black crow flew by and called out in a human voice:

'Ho there, young Alyosha Popovich! Do not rip open the Tatar's white breast, and I will fly to the blue sea and fetch the Waters of Life and Death. You shall anoint Dobrynia with the Water of Death and his white body will grow whole; you shall anoint him with the Water of Life and he will revive.'

Alyosha hearkened to the black crow, and it flew to the blue sea and brought back the Waters of Life and Death. Alyosha anointed Dobrynia with the Water of Death and his white body grew whole; he anointed him with the Water of Life and he awoke from his sleep of death, and they set the Tatar free.

And as the sun rose still higher in the heavens, Ilya of Murom rose and went down to the River Safat to wash. And barely had he dried himself on the thick linen towel and prayed to the holy icon, than he saw a great Musulman army crossing the river towards him, and seeing that he could not flee from the army even as a grey wolf or a black raven, he cried out in a loud voice:

'Where are ye, mighty Russian heroes? Awake from your slumbers and come to your brother's call!'

Hearing his call, the mighty heroes straightway sallied forth, leapt on their trusty steeds and threw themselves at the Musulman army, and cutting and carving among the hostile force for three hours and three minutes, they quite destroyed the pagan host.

And proud of their achievements, they began to boast.

'Our mighty shoulders are not weary, our trusty steeds are not tired, and our steel swords are not blunted, yet we have destroyed this great force!' exulted Vasili Buslayevich.

'Yea, though the angelic hosts of Heaven came down to fight us, brothers,' boasted Alyosha Popovich, 'we should not know defeat!'

And no sooner had he spoken these words than two unknown warriors appeared before them, and they cried out in a loud voice:

'Come on, ye Russian heroes, and fight with us. Take no heed that we are but two while ye are seven!'

Angered by this speech, the bold Alyosha lashed his fiery steed, and rushing upon the two warriors, he struck each of them in two with two fell strokes of his sword. But when he wheeled his horse round, he saw that the two had become four, and all four were alive. Then bold Dobrynia rushed upon them and with four swift blows he struck each one in two, and straightway the four became eight, and they were all alive. Then Ilya of Murom rushed upon them and with eight strokes of his sword he cut each one in two, and straightway the eight became sixteen, and they were all alive. Then all the heroes rushed desperately upon them and began to cut and carve, but the more blows they struck, the greater the enemy army grew. For three days the Russian heroes fought, for three days three hours and three minutes, and their mighty shoulders grew weary, their trusty steeds grew tired, their steel swords became blunt, and still the mighty force grew and grew, and advanced in frightening array upon the heroes. And for the first time the Russian heroes were frightened and took flight and ran to the stone mountains and the dark ravines to hide. But lo! as the first hero entered the hills, he straightway turned to stone; and as the second hero entered the hills, he too turned to stone, and the third, and the fourth, and every hero entered the hills and turned to stone, and from that day there have been no more heroes in the Holy Russian land.

FOLK-TALES

(Skazki)

Death and the Soldier

A LONG time ago in Russia there lived a Soldier who served the Tsar faithfully for the space of twenty-five years, fighting his battles on many fronts and receiving, alas! far more grievous wounds than golden coins. At the end of his twenty-fifth year in the Army, when he had grown old and was no good to fight any more, he was summoned by his captain.

'Soldier,' said the officer, 'you have served the Tsar faithfully for many years and have always performed your service well. But now that you are old the Army can no longer afford to keep you. You are therefore released from military service and may now go home. Private Ivan, dismiss!'

The Soldier saluted, turned about, slung his knapsack over his shoulder, and marched away smartly. But though his step was as brisk as ever, his heart was heavy with sadness.

'For twenty-five long years I have served the Tsar,' he thought, 'and now that I am no longer considered to be of any use to him, I am sent away with only three dry biscuits to stand between me and starvation. What can a poor old soldier do, and where shall he lay his head? There is nothing for it but to return to my native village, back to my old father and mother, and if they are no longer of the world of the living, I shall just lie down beside them in the grave and die.'

On and on he marched, and one day passed and then another until all that remained in his old battered knapsack was one lone biscuit. This he saved as long as his stomach could endure it, for his home was many miles away. When hunger rattled hollowly within him, however, and he could stand it no more, he sat down under a tree, took out the one dry biscuit, and was about to eat it when his eyes fell on a ragged old beggar standing before him.

'I am hungry, Soldier,' said the beggar. 'Give me charity, for the love of God.'

'Ah, well,' sighed the Soldier. 'I am an old sweat and still have a trick or two up my sleeve. Something is bound to turn up. But where should this poor old beggar turn if he is hungry?'

He handed the biscuit to the old man, lit his pipe, straightened his crumpled tunic, and set out once again along the road for home, marching on and on along the hard, dusty road with the hot sun burning down on him.

Suddenly he stopped dead in his tracks. A flock of wild geese were swimming on the surface of a lake.

'Here's a stroke of luck!' muttered the Soldier, and ducking swiftly among the bushes, he unslung his rifle, took careful aim, and with three quick and skilful shots knocked off as many geese.

'I shall not starve to death yet awhile,' he murmured as he picked the geese up and stuffed them in his empty rucksack, and striding gaily on into town, he came to a halt before an old ramshackle farm-house. He knocked loudly on the door, and when it opened and an aged peasant appeared in the doorway, he thrust the three geese into his hands.

'Take these three geese, batiushka,' he said. 'Roast one for me, keep the second for yourself as payment for your trouble, and in exchange for the third give me a goodly measure of wine to wash down my roast goose!'

That evening in the old farm-house the Soldier ate a meal fit for the Tsar, and the borsch, the roast goose,

and the strong wine made his belly roll with a contentment
it had not known for many a day. As he pushed his plate
away with a happy sigh and lit his old pipe, his eye chanced
to fall on a palatial mansion situated on the opposite side of
the road.

'That's a fine building, batiushka,' he said idly. 'To
whom does it belong?'

'That,' replied his host, filling two cups of tea from the
boiling samovar, 'was built by the richest merchant in
town.'

'He's a lucky fellow to have such a place to live in.'

'He *would* be,' replied the peasant, 'if only he could *live*
there. But he dare not move in.'

'Dare not move in?' repeated the Soldier incredulously.
'Why not?'

'Unclean spirits have seized the whole house. Every
night a host of demons gather in it and shriek and shout and
dance about until the light of dawn. All night long the place
is in an uproar and the folk who live in these parts are too
terrified to go near the place after dusk.'

'That is interesting,' said the Soldier. 'Where might I
find the owner of this house, for I should like to ask him
whether there is anything I could do to help?'

The peasant stared at the Soldier as if he had gone mad,
but reflecting that twenty-five years in the Army would
soften the brain of any man, he told him what he wanted to
know. The Soldier lay down on the stove to rest after his
fine dinner, and when evening came, he rose, went into
town, and sought out the rich merchant.

'What can I do for you, Soldier?' the merchant asked,
somewhat absently, for he was a worried man.

'I am on my way home,' replied the Soldier, 'and have
nowhere to spend the night. I should be very grateful if you
would allow me to sleep in your big new house down the
road, for I have observed that it is quite empty.'

'Are you mad?' exclaimed the merchant. 'It would be
suicide! Sleep anywhere else—for there are houses enough
in this town—but for Heaven's sake do not think of entering

that accursed new house of mine! As soon as I had the roof on, half the demons of Hell moved in. There is no getting rid of them.'

'Let's not look on the dark side of things,' said the Soldier. 'You never know, the demons might jump to it at the command of an old sweat like me.'

'Don't think you are the first to try to rid the house of this devil's brood, Soldier,' said the merchant. 'Others have undertaken the task, and all have failed. Last summer a fellow volunteered to drive them out of the house and had the courage to spend the night in it. All that remained of *him* the next morning was a heap of bones sucked white and dry!'

'We have a saying in the Army, merchant: "Water will not drown the Russian soldier nor fire burn him." I've served with the colours for most of my life, and many's the narrow escape I've had in countless battles and campaigns. But I've always come out alive and kicking and I should not like anyone to think that I should retreat or surrender in the face of a mere devil or so.'

'Have it your own way,' replied the merchant. 'If you are not afraid I shall not try to deter you. Indeed, if you do manage to drive these evil spirits from my house, I shall not be sparing with your reward.'

'All right,' said the Soldier. 'Now if you could get me a few candles, some dried nuts, and one large boiled beet-root . . .'

'Come into the store and take whatever you need,' said the merchant, and following him into the shop, the Soldier helped himself to a dozen candles, three pounds of dried nuts and one large boiled beetroot, and with this strange equipment he marched confidently off to the new mansion just as night was beginning to fall. He entered the house, made himself comfortable in the bedroom, hung his great-coat and knapsack on a hook, and lit the candles. Then he puffed away at his pipe, and cracking the dry nuts in his fingers, waited calmly and patiently for something to happen.

On the stroke of midnight a loud shriek of fiendish laughter echoed through the empty house, and this, it appeared, was the signal for a veritable bedlam to break loose, for on all sides and in every corner of the house doors slammed, walls rattled, ceilings shook, cloven hooves danced and clumped on the floorboards and blood-curdling screams rang through the air. The Soldier was puffing away at his pipe as though nothing had happened, when suddenly the door of the bedroom was pushed open, a red, horned head poked round the door, and the glowing green eyes of a devil lit up with demoniac glee.

'I've found a man! I've found a man!' he shrieked. 'Oh, what a feast we'll have now!'

There was a squeal of joy and a scurry of a thousand furry feet, and in a trice a host of devils crowded in the doorway, hustling and bustling and elbowing each other in their enthusiasm.

'Tear him to pieces and eat him up!' they shrieked.

'Just come and try it!' said the Soldier, with a threatening glare. 'I've never seen such a lousy crew in all my life! I've killed better devils than you a dozen at a time, so shut up, or else . . .'

In the face of this somewhat unexpected reception, the multitude of devils shrank back in some alarm. The largest of them was likewise daunted by the menacing tone of the Soldier, but urged on by his companions, he stepped forward with a show of bravado.

'Would you care to measure your strength against mine, human?' he sneered.

'Why not, indeed?' said the Soldier, taking his pipe from his mouth and rising slowly to his feet. 'Can you squeeze water out of a stone?'

The eldest devil ordered some stones to be brought up from the street, and a small devil was soon back with a handful. He gave one to the old devil and one to the Soldier.

'You try first,' said the Soldier.

The demon squeezed the stone so hard that it crumbled to dust.

'See that, human?' he exclaimed triumphantly.

The Soldier shrugged his shoulders contemptuously and took the large boiled beetroot out of his knapsack.

'This stone is far larger than those pebbles you brought up,' he said. 'Now look!'

He squeezed the beetroot lightly between his hands until the juice began to drip from it like drops of blood. The devils looked at one another in silent amazement. After an embarrassed pause in which only the sound of the Soldier's jaws was audible, they asked him what he was chewing.

'Walnuts,' replied the Soldier. 'But none of you will be able to chew these walnuts of mine.'

'Give me one and I'll try,' said the old devil.

Instead of a walnut, however, the Soldier threw him over a bullet from his rifle. The devil popped the leaden bullet into his mouth and chewed and chewed with all the strength of his jaws, but could not crush it between his teeth. The Soldier continued to put one walnut after another into his mouth and, of course, to chew them up quite easily.

The devils were confused and ashamed.

'Well,' said the Soldier, 'you don't appear to be very good at eating my walnuts, but I always heard that the likes of you were good at assuming all manner of shapes and sizes. I have been told that big devils can become little ones, and little devils big ones, and that you can squeeze yourselves into any crack or crevice.'

'So we can!' cried the devils in unison.

'I don't believe it!' said the Soldier, shaking his head. 'I wager you cannot all squeeze yourselves into my old knapsack.'

With that the demons rushed helter-skelter into the Soldier's knapsack, exulting at the chance to prove their superiority, and within seconds there was not one of them left anywhere in the whole house. When the last hairy tail disappeared into the knapsack, the Soldier strolled over, shut it up tight and strapped and buckled it firmly.

'Now perhaps we'll have some peace and quiet,' he muttered, and taking his overcoat off the hook, he wrapped himself up warm and slept soundly all night long.

The next morning the merchant sent two of his servants to find out whether the Soldier was still alive and in case he had perished at the hands of the demons, to bring back his bones. When they arrived, however, they found the Soldier wide awake and strolling round the room sucking his pipe.

'Good morning, Soldier,' they said. 'We certainly did not expect to find you among the living. Look, this box was for your bones!'

The Soldier laughed.

'It's a little early to think of burying me, I fear,' he said. 'But you can help me carry this knapsack to the blacksmith, if you like.'

They took the knapsack and carried it off to the smithy.

'Now, lads,' said the Soldier to the blacksmith and his mates. 'Put this bag on the anvil and hammer it flat for me.'

The blacksmith placed the bag on the anvil and all three of them beat on it with their heavy hammers until the sweat ran down their brows and the whole village resounded to their blows.

'Help! Help!' screamed the tormented demons. 'Soldier, have mercy! Let us out! Help!'

But the Soldier took no notice.

'Come on, lads,' he said to the smiths. 'Put some strength behind those blows! We have got to teach the likes of them not to take liberties with the likes of us!'

'Help! Mercy!' squealed the devils. 'We promise never to come near the town again, let alone the merchant's house, and we will give you a king's ransom if you will only spare our lives!'

'Very well,' said the Soldier. 'But remember! don't ever try any of your tricks on a Russian soldier again!'

He ordered the blacksmiths to cease hammering at the sack, undid the straps and buckles and let the demons out one by one until only their leader remained inside.

'First bring me his ransom,' said the Soldier, 'and then I shall release him also.'

Hardly had he time to light up his pipe than he saw a tiny devil approaching him carrying a leather bag.

'Here is the ransom,' squeaked the devil.

The Soldier took the bag and was surprised to feel that it was so light. He unfolded it, looked inside, shook it, turned it inside out, held it up to the sun, but still he could find nothing in it.

'Now that you have had your little joke,' he said, 'I shall let your chief feel the taste of the hammers all by himself!'

The old devil in the knapsack gave a terrified squeal.

'No! don't hit me, Soldier!' he cried. 'Listen! The bag is no ordinary bag. It is a magic one, and there is only one like it in the whole world. Make a wish, open the wallet and you will find that your wish has been fulfilled. If you want to capture anything or anybody, all you have to do is to open the bag, shake it and say "Come on in!" and no one addressed in this way will be able to resist.'

'I'll try it out to make sure it works all right,' said the Soldier, and thought to himself: 'What we could do with just now is three good bottles of beer.'

Immediately the bag felt heavy, and when the Soldier opened it he found three bottles of beer inside. He handed them round to the blacksmith and his mates, proposed their health, and walked outside the smithy. He looked about him this way and that, and spying a sparrow perched on a roof-top he opened the bag and shook it in front of him.

'Come on in!' he said to the sparrow, and hardly had the words left his lips than the sparrow flew down from the roof-top straight into the magic bag.

'Well, it works all right,' said the Soldier, 'and no doubt it will prove quite useful.'

And he went back into the forge and released the devil from his knapsack.

'But have a care,' he warned him grimly. 'If ever I set eyes on you or your brood again, I'll have you in this bag in a jiffy. And then look out for yourself!'

The devil ran off as fast as his legs would take him, his tail between his legs, and the Soldier went to the merchant.

'It's all clear now,' he said. 'You can move into your new house whenever you like.'

When the merchant saw the Soldier standing before him with his servants, his jaw dropped in amazement and he could hardly believe his eyes.

'By Heaven, what you said about the Russian soldier is right enough!' he cried. 'Water will not drown him nor fire burn him! How did you manage to get the better of the devils and come off scot-free?'

The Soldier told him all that had happened during the

night and how he had got rid of the devils for ever. The merchant's servants corroborated what he said, but the merchant thought it better to play for safety and to wait a while longer before moving into the house. In the evening, therefore, he sent his two servants to spend the night in the new house with the Soldier.

'Be on your guard,' he told them, 'and if anything happens, the Soldier will look after you.'

Nothing did happen, however, and they all passed a quiet and peaceful night in the house and returned safe and sound to their master the following morning.

The third night the merchant himself decided to sleep in the house, and again the night passed without incident. The merchant was overjoyed at his good fortune, and the next morning he ordered all his possessions to be transferred to his new house and a great banquet to be prepared. Everything of the best was cooked and boiled and roasted and baked, wine flowed in profusion and the tables creaked under the strain. The Soldier was given the place of honour at the head of the table and all the many guests ate and drank to their heart's desire.

'Eat, drink, and enjoy yourself, Soldier,' cried the host, 'for I shall never forget your services as long as I live!'

The merry company feasted until the early hours of the morning. When they finally broke up, the sun was already shining in at the window, and the Soldier hoisted his knapsack on his shoulder and prepared for the road.

The merchant was dismayed.

'What is the hurry, my good fellow?' he cried. 'Stay with us and be our guest for a week, or a month, or as long as you like!'

'I thank you for your kind offer, merchant,' replied the Soldier, 'but I must go home to see my parents. I was only too glad to be of assistance to you.'

The merchant opened his money-coffers and was about to fill the Soldier's knapsack with silver coins. The Soldier declined politely.

'You are very kind, merchant,' he said, 'but I have no use

for silver. I still know a trick or two and shall no doubt get on all right by myself.'

And taking his leave of the merchant, he stuffed his magic pouch into his knapsack and marched boldly out into the world. For a long way or for a little way he marched on until he came to a hill, and when he had climbed to the brow of the hill and looked down and saw the village where he was born nestling in the valley below, his heart was filled with joy and his soul felt as light as a feather. He quickened his pace and marched eagerly down the hill-side, gazing all the time around him at the budding trees, the silver lakes, the green hills, and the golden sun and sighing with happiness and pride.

'How beautiful you are, my Russian land,' he murmured. 'Many are the lands I know, and rich and wonderful are their towns, but none in the whole white world is as fair as you!'

And he walked joyously up to the little hut where he was born and knocked on the door. A bent old woman appeared on the threshold and the Soldier threw himself into her arms.

'My son, my son,' cried the old woman. 'We had long since given you up for dead. Your father longed to give you his blessing before he died, but alas! it is five years since he passed away with his wish unfulfilled.'

The Soldier's old mother began to weep, but he kissed her cheek and hugged her and told her that she would never have to worry any more.

'Don't fret, matushka,' he said softly. 'Now that I am here I shall look after you.'

He undid his pouch, thought of all manner of good things to eat and drink, and the magic pouch grew fat and bulky. The Soldier emptied it on the table.

'Eat and drink to your heart's content, matushka,' he said, but the old woman was so overcome with emotion that she could scarcely touch a thing. But for all that she was very happy.

The next day the Soldier again undid the magic pouch, wished for some silver, and began to put the affairs of the farm in order. He built a new cottage, bought a cow and pigs and sheep in the market-place, and within a few weeks he had become one of the most prosperous farmers in the province. Then he found himself a bride, married and had children, and his old mother, whose happiness knew no bounds, looked after her grandchildren with great care.

For a long time they lived in perfect peace and happiness, but when six or seven years had passed, the Soldier suddenly fell ill. One day went by, then another, then a third, and still he could not rise from his bed and eat and drink. He grew weaker and weaker and then, on the eve of the fourth day, opening his eyes, which felt as heavy as lead, he saw that someone was standing at his bedside. It was a white skeleton! Slowly and deliberately the horrible vision whetted its great scythe and stared grimly at the sick man. It was Death!

'I've come for you, Soldier,' he said. 'Get ready!'

'Wait a little longer, Death!' protested the Soldier. 'There's still a good thirty years left in me! I've got to raise my children and marry them off. Come and see me again when I've done that, for I have no time to die at present.'

'There's no leave in *my* Army, Soldier,' chuckled Death, his bones rattling drily. 'You won't get thirty years or thirty minutes from me.'

'If you won't give me thirty years, Death,' pleaded the Soldier, 'give me three years' grace! You know I've a lot to do.'

'Not three years nor three minutes!' exclaimed Death, impatiently sharpening his scythe. 'You've got your marching orders, so jump to it!'

'*You* jump to it!' said the Soldier, and summoning up all his strength he pulled his magic pouch from beneath his pillow and waved it with a feeble hand. 'Come in here!'

Scarcely had the words left the Soldier's lips than Death became a struggling bag of bones. The Soldier felt as though a heavy weight had lifted from his chest and he knew that he was better. He fastened the pouch securely, got up, sat at the table and ate a substantial meal of wholesome black bread, white cheese, and brown beer. When he finally pushed back his plate he felt as strong as ever, and sauntering up to the magic pouch he said:

'Well, Bony, you were not very civil to me, were you? That will teach you not to try to pull a fast one on a Russian soldier!'

A small, thin voice emerged from the depths of the bag.

'What are you going to do with me, Soldier?' asked Death.

'Well, I did not really want to lose a good pouch,' said the Soldier gravely, 'but there is nothing for it. I am going to drop you into the swamp and push you well down into the mud. You won't get up to any of your mischief there, I'll be bound!'

'Let me out, Soldier!' squeaked Death. 'I'll give you three years' grace!'

'What do you take me for? I've got you now, and I intend to keep you!'

'Let me out. I'll make it thirty years!'

The Soldier rubbed his hand thoughtfully over his stubbly chin.

'That's more like it,' he said. 'But if I let you out, you must promise me not to meddle with people's lives for thirty years, so that no one shall die in all that time.'

'Soldier, I cannot promise that! What shall I live on?'

'You can eat roots and radishes in the fields.'

Death made no answer, and the Soldier pulled on his boots.

'Have it your own way, Death,' he said. 'Hold tight, we're off to the swamp!'

He swung the bag roughly over his shoulder.

'Stop!' cried Death. 'You win, Soldier. Let me go and I promise to live on roots and radishes and not to kill anyone for thirty years!'

A stifled sob sounded from the depths of the bag.

'All right, then!' said the Soldier. 'But no treachery, mind! Keep your promise!'

With that he let Death out of the bag. Death jumped out, straightened his bones, picked up his scythe, and rattled off into the wood as fast as his legs would carry him. There he fell to eating roots and radishes. The Soldier's good flesh would have been better, but alas! there was nothing to be done.

And so the Soldier lived happily on with his mother and wife and children with never a thought for Death. One day, however, when he was busy sowing in the fields, he felt a sudden tap on his shoulder, and with some surprise saw the grinning skull of Death beside him.

'Your thirty years are up, Soldier,' he said. 'Are you ready?'

'Ah, well!' said the Soldier, shrugging his shoulders. 'A deal's a deal, and discipline's discipline. If my time is up, bring me my coffin.'

Death required no second bidding. He gleefully dragged up a large coffin with wrought-iron clasps and raised the lid.

'Hop in, Soldier!' he said.

The Soldier's face flushed in anger.

'Hop in? *Hop in?*' he exclaimed. 'How do you think I know the drill for getting into a coffin? I am an old soldier. You show me the movements and I'll perform the manœuvre in correct order.'

'Very well, Soldier,' said Death. 'One might as well die as one has lived. I'll show you what to do.' He laid himself carefully into the coffin. 'Look, it's easy. Just lie down naturally, so. Stretch out your legs, so. Then cross your hands at right angles on your chest, so. One, two, three, so!'

The Soldier, however, had always been better on the battlefield than the parade-ground, and as soon as Death was comfortably settled in the coffin, he slammed down the lid, banged in the nails, loaded it on to a cart, and drove it off to the river.

'You go, Death!' he chuckled. 'I don't think I shall volunteer after all.'

With a great splash, the heavy oak coffin fell into the river, and the swift stream bore it rapidly out to sea, where for many years Death floated helplessly around on the choppy waves.

When some years had passed, however, and Death's bones were beginning to turn a little green, and seaweed and barnacles covered the coffin, a severe storm blew up, huge waves dashed against its sides and finally swept it hard against the rocks on the seashore. The wood splintered and shattered, and Death, more dead than alive, flopped wearily on to the beach like a stringless marionette. He rested there for several hours, cleaning his rusty scythe and muttering to himself, and then rose to his feet and set out in search of the wily old Russian soldier. By now he had learned to be cautious in his dealings with him, and when he came to his farm, he cunningly hid in a corner and waited patiently for him to pass by.

The Soldier was preparing to go and sow the corn at the time, and he was just going into the granary with an empty sack to fill with seeds when Death darted out from behind the door and stood before him.

'Got you!' he cried. 'Surrender, Soldier, for you won't get away now!'

'I'm in a tight corner here!' thought the Soldier, but taking his empty and harmless corn sack from his breast, he waved it threateningly in Death's grinning face.

'So you're back again, are you?' he cried. 'I take it you want to spend a holiday in my sack at the bottom of the swamp?'

As soon as Death saw the empty corn sack waving in his face, he took it to be the Soldier's magic pouch, and though the grin could not exactly disappear from his fleshless face, he turned as pale as death is usually thought to be, and before the Soldier could say another word, he took to his heels and fled, if one may use the expression in his case, for dear life.

And thus it was, they say, that the Soldier was the very last person ever to see Death clearly in front of his eyes, for since that day Death always creeps up on people unseen and invisible. His one preoccupation is not to be seen by the Soldier, for there is nothing he dreads so much as spending eternity at the bottom of the slimy green swamp. And the Soldier lived happily ever after, and even now he is still alive and kicking.

That, at least, is how some say the story ended, but there are others who intimate that the Soldier's victory over Death was not so clever, after all, and had quite serious consequences for himself and his fellow men. These people say that when the Soldier first trapped Death in his bag he did not release him again in exchange for any promise, but took him into the dark Brianski Wood, hung him on the top of a trembling aspen-tree, and returned home intending to get on with his work unimpaired by the importunity of this wearisome fellow Death. And from that day on, although baby after baby was born as regularly as before, no one in the whole white world died at all.

Many years passed by and the Soldier had all but forgotten about poor Death hanging up in a bag on the top of a tree, when one day, on his way to town, he met an old, old woman coming along the road in the opposite direction, and the old woman was so weak and frail and her great years weighed so heavily upon her poor bony shoulders that she blew from side to side at the slightest gust of wind.

'Lord! What an old woman that is, to be sure!' said the Soldier. 'She ought to have been dead long ago, poor old soul!'

'You are right, Soldier,' quavered the old woman. 'I ought to have been dead long ago. When you tricked Death to enter your bag these many years ago, I had but one more hour to live. I should be glad to go to my rest, Soldier, for Death is merciful to such as me. Beware, Soldier, for the day will come when you too will suffer what you have inflicted on me!'

With this, the old woman went her way and left the Soldier deep in thought.

'The Russian proverb says that old age is no joy but death is no gain,' he thought to himself. 'But the old woman is right. I have already many sins against my name and am already pretty old. It is better to die now and avoid the torments of a decrepit old age unworthy of a soldier. And the older I get, the more I'll suffer.'

So the Soldier went to Brianski Wood and found the aspen-tree at the top of which the bag was rocking dizzily to and fro in the breeze.

'Are you still alive, Death?' called the Soldier.

'Only just, Soldier,' faltered a weak and muffled voice from inside the sack.

The Soldier climbed the tree, took down the bag, and carried it home. Then he lay down on the bed, bade farewell to his wife and children, and opening the bag, asked Death to finish him off.

As soon as the bag was open, however, Death leapt to the floor, and out of the door in a flash, he made off as fast as his legs would carry him.

'Go to Hell and may the devil take you, Soldier!' he called back. 'I never want to set eyes on you again!'

With that he was gone, and there was nothing for the Soldier to do but get up and get dressed, for he was as alive and well as ever. The thought that he would grow older and older and never die, however, began to prey on his mind, and in the end he decided to follow Death's parting advice and persuade the devils to take him.

'They will throw me into a cauldron of boiling pitch,' he thought, 'but in the end my sins will be washed away.'

Taking once again his leave of his wife and children, he walked down to Hell with his magic pouch in his hand until he came to the grim sentinel at Hell's Gate.

'Who goes there?' challenged the devil.

'A sinner who has come to be tormented for his sins,' replied the Soldier.

'Come in,' said the devil. 'We don't have many volunteers for our Army. Wait a minute, though! What is that in your hand?'

'A sack.'

The sentinel took a sharp pace to the rear.

'Help! Sound the alarm!' he yelled.

The devils scurried to and fro and within seconds the Gates of Hell were barred and bolted and the windows shuttered. The Soldier scratched his head, for he had not expected this unfriendly reception, and walking round the walls he called out to the Prince of the Devils.

'Let me into Hell, Beelzebub!' he cried. 'I have come to be tormented for my sins.'

'There's no room here for you,' said Beelzebub. 'It wouldn't be long before we were all struggling and suffocating in that blessed bag of yours!'

'Very well,' said the Soldier. 'If you won't let me in, at least give me two hundred souls that I might lead them to Heaven. Perhaps God will forgive me my sins for this deed.'

'You can have two hundred and fifty souls,' cried Beelzebub, 'but go away from here!'

The Prince of Hell ordered two hundred and fifty souls to be let out at the back so that the Soldier should not somehow slip inside, and the Soldier formed up his platoon in fours and marched them off to Paradise.

When the Apostles saw the band approaching, they informed the Lord and asked for His instructions.

'Let the two hundred and fifty souls pass through the Gates of Heaven,' said God. 'But for goodness sake don't let the Soldier in, or he'll have us all inside that infernal bag of his before long.'

When he heard this bad news, the Soldier took one of the souls aside, and giving him his magic pouch, he whispered:

'Once you're inside the Heavenly Gates, just open the bag and say aloud: "Soldier, jump in here!"'

The soul took the bag, the Heavenly Gates opened, and the two hundred and fifty souls marched in. Such was their joy upon entering Heaven, however, that the soul forgot all about the Soldier and his bag, and after waiting a long time in vain outside the walls of Paradise, the Soldier resigned himself to his fate and walked sadly away to wander unhappily about the wide, white world for all eternity, unwanted and unwelcome in every place.

And only the other day, they say, did he finally die.

I-Know-Not-What of I-Know-Not-Where

LONG, long ago in Old Russia when Baba Yaga, the dreaded old witch of the forest, was but a few thousand years old and animals and birds and trees still deigned to speak to Man in his own language, there lived a mighty Tsar who was so exceedingly rich that he had everything he could possibly wish for. Everything, that is, except for one thing, for the Tsar had no wife. The old Russian grandmothers would shake their heads sadly and wonder what the world was coming to, while the old grandfathers would wink to each other and cackle among themselves.

'Not only is the Tsar rich and powerful,' they would whisper, 'but he is exceedingly wise to boot.'

Now you must know too that this Tsar had a young Archer by the name of Petrushka whose well-aimed bow regularly supplied the choicest morsels of game for the King's table. On the day our story begins, however, Petrushka was beginning to think that his luck had quite deserted him. He had tramped about the fields and forests of the Tsar's domains all day long with not a single bird or beast to show for it, and as evening fell, he trudged wearily home with his empty sack on his back. His heart was heavy, for he was apprehensive of the Tsar's impending wrath and felt that his head now rested but insecurely on his shoulders. Suddenly, however, he stopped in his tracks. There before him, perched peacefully on a tree, a white dove sat preening its feathers.

Beauty is beauty, thought Petrushka, but business is business, and rapidly bringing up his bow, he drew back the string and loosed an arrow at the bird. The steel tip of the arrow pierced the dove's right wing, and with a cry of despair, the bird fell wounded to the ground, a trickle of blood staining its white breast. Petrushka ran to where the bird had fallen and was about to wring its neck in triumph when it suddenly cried out in a human voice.

'Please don't kill me, Archer!' it pleaded. 'Take me home with you, dress my wound, and allow me to rest a little. When you see a drowsiness creep upon me, strike me with your right finger, and you shall reap a rich reward.'

Ah well, thought Petrushka, the Tsar will have to make do with common fare tomorrow; and carrying the bird gently home he dressed its wound and placed it on the window-sill to rest. After a few moments, the bird's eyes drooped as the soft heaviness of slumber weighed upon them, and tucking its tiny head beneath its one sound wing it prepared to sleep. Then, remembering what the dove had told him to do, Petrushka softly tapped the gently stirring bunch of white feathers with his right finger. And then something wonderful happened! The white dove fell from the sill, and before Petrushka's astonished eyes, it turned into the most beautiful maiden that ever graced a Russian folk-tale. Her face was as round as the moon, her hair as golden as the sun, her eyes as blue as the deep sea, and her skin as clear as mother-of-pearl, and before the Archer could recover from his surprise and shake the rosy clouds of wonderment from his head, the damsel spoke. And her voice was as soft and caressing as that of a turtle-dove.

'Archer,' she said, 'you have succeeded in catching me, and now you must contrive to keep me. Let us now be married, and I shall always be a true and faithful wife to you.'

Does anyone doubt whether Petrushka accepted? If so, let him doubt no more, for the Archer's assent was delayed only by the effects of his great delight, and as soon as his tongue could speak and his limbs move, he stammered his most willing consent and took the fair maiden in his arms. And so it was that the Archer married the Dove Maiden and they lived happily together—no, not yet for ever after; that would be too simple, for true happiness, in Russian folktales as in life itself, is not so easy to achieve. There were troubled times to come.

Although they were very happy together, their life was not an easy one, and one evening, when Petrushka returned tired and weary from a day's hunting with but little to show in his sack, Masha—for that was the Dove Maiden's real name—laid her hand on her husband's shoulder and said:

'We are very poor, Petrushka.'

'That's true,' sighed the Archer sadly. 'There's no denying that.'

'Have you got a hundred roubles?' asked Masha, and her husband replied that he had indeed a hundred roubles, but that it represented all the wealth he possessed.

'Well, then,' continued his wife, 'take your hundred roubles, go to town, buy me a hundred different skeins of silk, and I will make us rich.'

Well, nobody can say a Russian is averse to a gamble, and following his wife's instructions, Petrushka went forthwith to town and soon returned with a hundred skeins of the finest silk.

'Go to bed and rest now,' said Masha, 'for morning is wiser than evening.'

As soon as Petrushka had climbed upon the stove and fallen asleep, she took her hundred skeins of silk, sat down on a stool, and began to weave by the light of a candle. All night long her clever fingers worked, and when her husband awoke the next day, she held up the carpet in the pale light of dawn, and it shone like the sun. There never was such a carpet in Persia, Caucasus, or Hind! All the Kingdom of Russia, from the Baltic to the Urals, from the White Sea to the Black Sea, was woven therein, and the trees seemed to rustle, the birds to sing, the beasts to growl, the fish to leap,

the soldiers to march and the ships to sail, and the sun and
the moon and the stars all shone in the silken sky.

'Take this carpet to the merchants in the market-place,'
said Masha. 'Put no price on it yourself, but sell it to the
first man to make you an offer.'

The Archer rolled up the carpet, and tucking it under his
arm went to the market and spread it on the ground for all
to see. The merchants, Russian, Jewish, Armenian, and
Greek, all ran quickly up to see what wonder had been un-
furled before their eyes.

'How much?' said one.

'Name a price,' said Petrushka, and the merchant
scratched the back of his head and thought and thought,
but could not name a price for the carpet. And all the
merchants stood around and marvelled, but none could guess
a fitting price for the carpet. Just at this time, however,
a counsellor of the Tsar chanced to ride through the
market-place on his way to the palace, and seeing the crowd
of merchants thronging around something in a most un-
usual silence, he rode up to them and dismounted.

'Greetings, merchants,' he said. 'What is happening?'

Then he saw the carpet.

'Gospodi!' murmured the counsellor. 'Where is the
human hand that could create such a marvel?'

'My wife wove it,' said Petrushka, stepping forward.

'What is its price?' asked the counsellor.

'I do not know myself. I am to accept the first offer.'

Now the merchants had not dared to quote a price for
fear of seeming ridiculous, but who would expect a State
official to be as sensitive as that?

'I'll give you five thousand roubles for it,' said the Tsar's
minister.

'You have named a price,' said the Archer. 'The carpet
is yours.'

And handing it to the counsellor, he put the five thousand
roubles in his sack and went home, while the counsellor rode
proudly back to the palace and displayed his new acquisition
to his master.

'Why, it's a picture of my Kingdom!' exclaimed the Tsar. 'I must have it! How much do you want for it?'

'It cost me a lot of money,' said the counsellor, looking worried and rubbing his chin. 'Fifty thousand roubles would not be excessive.'

'Done!' said the Tsar, and paying the minister the price he asked, he retired to his private chambers to admire the wonderful carpet alone.

The counsellor saw that there was money to be made in the carpet business, and deciding to approach the Archer's wife and order another carpet similar to the first, he galloped out of the palace and rode up to the lowly hut where Petrushka lived with his wife.

He knocked on the door, the door opened and the Dove Maiden stood in the doorway. As soon as the counsellor's eyes fell on her radiant beauty, his senses whirled and he had to hold on to the wall to prevent himself falling down from sheer ecstasy.

'Yes?' said Masha.

But the counsellor's tongue was completely paralysed and he could not answer, and thinking the man was a fool, Masha shrugged her shoulders, shut the door, and went back to her work. The vision gone, the counsellor came slowly back to his senses and trudged sorrowfully back to the palace.

'What is the matter with you?' said the Tsar, staring at his minister's woebegone countenance.

'Sire, I have just been to the house of your Archer, and I swear his wife is so fair that in all Russia there is not a maiden to equal her beauty!'

'Ah? So?' thought the Tsar. 'If she's as beautiful as he says, I must see her for myself.'

And so, disguising himself in everyday clothes, the Tsar walked to the lowly hut where Petrushka lived with his wife and knocked imperiously at the door. The door opened and the Dove Maiden stood on the threshold.

'What do you want?' she asked, but from the moment the Tsar's eyes fell on her beautiful countenance, his senses whirled and he too was obliged to hold on to the wall to prevent himself falling. His tongue was likewise paralysed and he could not answer, so Masha again shrugged her shoulders and shut the door. The vision gone, the Tsar came slowly to his senses and walked sadly back to the palace.

'Verily,' he said, 'this is a woman fit to be my Queen. She is much too good for a mere archer.'

For days and days the image of the beautiful Dove Maiden filled his mind and he could think of nothing but how to steal her away from her husband. Finally, he summoned his counsellor, the same one who had bought the carpet.

'Find me a way to get rid of that archer fellow,' he said grimly. 'If you succeed, you will be well rewarded with towns and villages without number and a ton of gold, but if you fail, your head shall surely fall from your shoulders!'

Thus encouraged and inspired, the counsellor went away to try to devise a scheme to dispose of Petrushka's undesirable presence in the world of the living, but think as he might, he could devise nothing that stood any chance of success, and finally, in desperation, he entered the royal pothouse to drown his sorrows.

'Why the hangdog look, brother?' called out a drunken muzhik as he entered, slapping the counsellor on the back.

'Leave me alone, drunkard!' snapped the counsellor. 'I've got a lot to think about.'

'Give me a goblet of green wine to drink,' persisted the drunkard, 'and I will help you to think.'

The counsellor bought the fellow a goblet of wine, and more in search of solace than hope of aid, he opened his heart to the drunkard and told him of the task the Tsar had given him.

'Your troubles are over!' cried the muzhik. 'I know Petrushka the Archer very well. He's only a simple fellow. All you have to do is to set him an impossible task and you'll soon see the last of him. Just tell the Tsar to send him to

find out how the late Tsar his father is faring in the other
world, and you can be sure the Archer will never return
again.'

'That's good,' said the Tsar, rubbing his hands gleefully
when he heard of the scheme.

'That's bad,' thought the Archer when he heard the
Tsar's command and learned that his head should fall if he
did not succeed in this quest.

'What is wrong?' asked Masha when Petrushka returned
home and sat staring glumly at the floor. Petrushka told her
everything and was surprised to see that his wife did not
share his gloom.

'That's nothing,' she cried. 'Go to the Tsar and ask him
to give you his counsellor to be your companion, for other-
wise he will not believe you whatever you say. Then take
this golden ring, cast it on the ground in front of you and it
will lead you straight to the dead father of the Tsar.'

The Tsar granted Petrushka's request and gave him the
counsellor to take along with him, and once outside the city
walls, the Archer cast the golden ring in front of him, whereupon it began to roll purposefully on across field, river,
swamp, and mountain with the Archer and the counsellor
following in its wake. Did they go a long way or a short
way? We do not know, but in any case the golden ring
eventually rolled into a dark forest and then on down into a
dark, deep pit. The earth smelled damp and strange
shadows flitted eerily among the leafless bushes, and the
Archer and the courtier shuddered. Then——

'Look!' gasped the counsellor, and he clutched the
Archer's arm in great fear.

There before them in the dim subterranean light an old,
old man with white hair and a long white beard was pulling
a heavy cart laden high with wood, while on either side of
him two red, horned devils belaboured his old bones with
heavy clubs and urged him on to greater efforts.

'Is not that the late Tsar?' whispered Petrushka.

'Yes,' replied his companion. 'That is he.'

'My lord devils!' cried Petrushka, stepping up to the

small group. 'I beg you to free this lost soul for a few minutes that I may talk a while with him.'

'It's all very well to ask us to free him,' said the devils, 'but who will do the work?'

Petrushka gave the courtier a shove, and he stumbled forward.

'Here's a strong live man to take his place for a little while,' said Petrushka, and straightway releasing the old man, the devils seized the counsellor and harnessed him to the heavy cart in a trice.

'Pull hard! Pull hard!' shrieked the devils, and beating their new servant with their wooden clubs, they drove him off into the woods.

Turning to the old man, the Archer told him how he had been sent to the nether world by the present Tsar his son.

'He wishes to know how you fare down here,' he said.

'Alas, good Archer,' sighed the old Tsar. 'Heavy and grievous is my lot in Hell. Give my son my greetings and tell him I advise him to be less harsh and cruel in his dealings with his subjects than I was, or he shall surely share my most unenviable fate.'

When the devils returned with the empty cart they released the counsellor and harnessed the old man in his place. Petrushka bade the late Tsar farewell and the two travellers returned the way they came.

When the Tsar saw the Archer, he was beside himself with rage.

'How dare you return here with your mission unaccomplished?' he shouted.

'My mission is accomplished, Sire,' said the Archer. 'I return from the other world with the greetings of your old father, the late Tsar. His life after death is extremely hard and he asks you to be less harsh and cruel in your dealings with your subjects than he was, for fear you too might share his fate.'

'Liar! Rogue! How dare you speak to me like that?' screamed the Tsar. 'How can you prove you have been to the other world?'

'Sire, if you look at your counsellor's back, you will see the weals made by the devil's heavy clubs when he took the place of your father for but a few moments.'

The Tsar angrily tore the shirt from the counsellor's back, and seeing the marks of a severe beating, he was compelled with extreme reluctance to recognize the veracity of the Archer's report, and had to dismiss him with his head still safely on his shoulders.

When the latter had gone from the palace to rejoin his wife, however, the Tsar turned furiously on his unfortunate counsellor.

'If you don't devise a certain way to get rid of this Archer,' he snarled, 'I'll chop off your head with my own hands!'

The counsellor thought and thought, but he could devise nothing, and finally he trudged wearily to the royal pot-house to drown his sorrows.

'What? Still in trouble?' called out the drunken muzhik who had given him advice the last time. 'Order me another goblet of green wine, and I'll try again to help you.'

When the counsellor told him of this present predicament,

the drunkard nodded his head and craftily winked his blood-shot eye.

'Go back to the Tsar, brother,' he said, 'and tell him to send the Archer to the Thrice-Ninth Kingdom in the Thrice-Ninth Land, and order him to bring back Kot Bayun, the Talking Cat.'

'That's good,' said the Tsar, when he heard of the scheme.

'That's bad,' said the Archer.

'That's nothing,' said Masha. 'Go to sleep on the stove and wait till tomorrow, for morning is wiser than evening.'

When Petrushka awoke the following morning his wife gave him three felt hats, a pair of pincers, and three metal rods.

'Go to the Thrice-Ninth Land,' she said, 'to the Thrice-Ninth Kingdom, but as soon as you have set foot there be on your guard. The Talking Cat will endeavour to overpower you with heavy slumber, but you must keep going and on no account give way. Put one foot before the other although they may feel as heavy as lead, and when you can walk no more, crawl on your hands and knees. If you go to sleep, you are lost, for the Cat will tear you to shreds.'

The Archer listened carefully to his wife's instructions and set off on his strange quest. Did he go a long way or a short way? We do not know, but at all events he walked on until he saw before him, veiled in a blue haze in a misty valley, the Thrice-Ninth Kingdom in the Thrice-Ninth Land. He had not advanced more than a few versts into this mysterious country when a heavy drowsiness fell upon him and he felt a huge weight hang on his eyelids and his limbs grew as heavy as lead. But he put the three hats on his head and dragged slowly and painfully on, and when he could walk no more, he crawled steadfastly on on his hands and knees. Suddenly the great weight of slumber fell from his body, and rubbing his eyes, he saw before his nose a great pole rising high up into the sky, and before he had time to look up a huge, black furry shape sprang down from the pole with a hiss of fury, and flattening its ears and narrowing its green eyes, the Talking Cat clawed viciously away at the Archer's head. The first hat and then the second were torn to shreds, and already the third was flying in pieces into the air, when Petrushka caught the Cat's left leg in his strong pincers and began to belabour him with the first of his metal bars. When the iron bar broke, he seized the bar of brass, and when the bar of brass broke, he seized the bar of tin, and though it bent, the tin bar did not break, but thudded relentlessly against the Cat's furry ribs with all the force that the Archer could summon into his arm.

'Help! Help! Stop it!' screamed the Cat. 'Stop beating me and I'll tell you all my best stories—stories of Kings and Queens and Popes and Deacons and the Sons and Daughters of Priests!'

'Keep your stories to yourself, animal,' the Archer panted, beating away as hard as ever.

'Stop! Stop!' cried the Cat, twisting and turning like a soul in Hell. 'I can stand no more! What do you want of me?'

'You must promise me to leave the Thrice-Ninth Kingdom in the Thrice-Ninth Land and return with me to the Tsar!'

'Anything! I'll do anything!' sobbed the Cat. 'Only stop your beating !'

'Do you promise solemnly?' asked Petrushka.

'I promise!' yelled the Cat, and to tell the truth, it was with great relief that the Archer ceased to hit the Cat, for his arm was tired and the tin had grown so hot that his hand was sore and blistered.

And so the Archer returned to the Tsar leading the Talking Cat behind him.

'I have accomplished my mission, O Tsar,' said the Archer. 'Here is the Talking Cat.'

'Oho!' exclaimed the Tsar, forgetting for a moment in his curiosity his anger at seeing the Archer back. 'Let him go and show us what he can do.'

But no sooner had he said this than the great Cat humped its back, bared its teeth, switched its tail, and with a hiss and a snarl it leaped straight at the Tsar's throat.

'Help!' shrieked His Majesty. 'Archer! Keep it away!'

With this the frightened Tsar tripped over his scarlet robes and fell in a heap on the floor, and he would have proved an easy prey for the ferocious Talking Cat had not Petrushka forced it to return by brandishing his pincers and tin club. The Cat sprang obediently back to the Archer's side, and taking it away and locking it up in a cage in the palace, the Archer returned thankfully home to his fair young wife.

What with the painful scratches round his throat and his thwarted plans, the Tsar was beside himself with rage.

'If you do not this time find out a sure way to rid me of this accursed Archer once and for all,' he bellowed at the crestfallen counsellor, 'your head shall surely roll!'

Not pausing to waste time on fruitless reflection, the counsellor made straight for the royal pot-house as fast as his legs would take him.

'What? Still in trouble?' said the drunkard, staggering out of his corner. 'This Archer is a sly one, after all!'

And sipping at the goblet of green wine the counsellor bought him, the crafty muzhik knitted his brows and scratched the back of his head.

'Now listen carefully, Your Excellency,' he said at length. 'All you have to do is to tell the Tsar to send the Archer to I-Know-Not-Where with instructions to bring back I-Know-Not-What. If that doesn't get rid of him once and for all, nothing will!'

'That's good!' said the Tsar.

'That's bad!' thought the Archer.

'Don't worry, Petrushka,' said the Dove Maiden. 'Go to sleep now, for the morning is wiser than evening.'

This time, however, the Dove Maiden did not know what to do herself, and though she stayed up all night racking her brains, and though she consulted all the books on magic at her disposal, she could think of nothing. An hour before the break of dawn, therefore, she went wearily to the window and waved her handkerchief.

'Birds of the air and beasts of the field!' she called. 'Come to my aid!'

Immediately all the birds of the air and the beasts of the field came to the lowly hut and sat beneath the window.

'Birds of the air and beasts of the field,' said the Dove Maiden. 'Know ye aught of the land of I-Know-Not-Where and the thing I-Know-Not-What?'

But the birds and beasts looked at each other dumbly, and even the wise old owl and the wily old fox were at a loss.

'Nay, Dove Maiden,' they replied at last, 'we know nothing of such a place or thing.'

Masha waved her handkerchief again and the birds of the air and the beasts of the field returned whence they had come. Then she went to the door, and standing on the threshold, she once more waved her handkerchief, and straightway there appeared before her two enormous giants.

'Pick me up, good giants,' she said, 'and place me down in the middle of the ocean.'

One of the giants stooped down and picked her up, and stepping into the middle of the great ocean, he held her on the palm of his hand on a level with the surface of the blue waters.

'Fish and reptiles of the deep blue sea!' cried the Dove Maiden, and the waves surged and rocked as a million silver forms swarmed around the legs of the great giants. 'Know ye aught of the land of I-Know-Not-Where and the thing I-Know-Not-What?'

But the fish looked at each other, and even the wise old whale shook its enormous head in regret.

'Nay, Dove Maiden,' replied the fish and reptiles of the deep blue sea. 'We know nothing of such a place or thing.'

Masha brushed her brow wearily with her hand.

'Take me back home, good giants,' she sighed, and stepping out of the ocean, the giants put her down on the threshold of her lowly hut, and disappeared. The golden dawn was just breaking over the dark forest when Masha woke her husband.

'Take this ball of wool and this embroidered towel, Petrushka,' she said. 'Throw the ball of wool in front of your feet and follow it wherever it leads. Where it stops, stop you also, and wherever it may be, you must find an opportunity of washing yourself and drying your body on this embroidered towel.'

Petrushka got ready for his journey and set off. He cast the ball of wool in front of him and it began to roll away neither too fast nor too slow. The ball rolled on and on, and the Archer followed it until his eyes ached and his head swam from watching it.

Whether he went a long way or a short way we no longer know, but eventually, just as he stopped to mop his brow in the middle of a dark wood, he saw to his surprise that the ball of wool had come to rest at the foot of a tiny hut which was supported on chickens' legs. It was the hut of the old witch Baba Yaga!

Petrushka surveyed it curiously for some minutes. He had heard of its existence but had never yet had the fortune or misfortune to see it for himself. It was evident that it was here that he was meant to stop, however, and there was nothing for it but to go on in.

'Izba, izba!' called out the Archer, who had learned the formula from his grandmother. 'Turn thy door to me!'

And immediately the hut span round on its yellow legs and the Archer saw a wiry old woman sitting inside spinning at her wheel.

As soon as the hut span round, however, the old hag stopped her spinning, and began to sniff the air.

'Faugh! Faugh! Faugh!' cackled the witch gleefully, jumping up from her spinning-wheel with surprising agility.

'Naught did I hear, naught did I see,
But Russian blood is come for me!'

Then she spied Petrushka standing before her on the threshold.

'Ah, there you are, my fine fellow!' she said, seizing the Archer with her brown and scraggy arm. 'I shall roast your flesh for supper and suck your lily-white marrow-bones dry!'

'Wait, Baba Yaga!' said Petrushka. 'You wouldn't like the taste of a dust-begrimed, travel-worn man. Heat up the bath as well as the oven, and by the time you are ready to cook me, I shall have washed my body clean and shall be fit to eat.'

Baba Yaga cackled gaily to herself, delighted to have found such an obliging victim, and she heated up the bath as well as the stove. Petrushka washed himself and then began to wipe himself dry on the Dove Maiden's embroidered towel.

'Where did you get that from?' asked Baba Yaga, eyeing him suspiciously. 'I recognize that needlework!'

'My wife made it for me,' replied the Archer.

'Then your wife must be my daughter!' exclaimed Baba Yaga. 'Welcome, my beloved son-in-law!'

And excusing herself profusely for her inhospitable behaviour, she prepared for him a fine meal with the choicest of meats and the sweetest cakes and all manner of beer and wines, and as they dined together, the Archer told his mother-in-law of his strange quest.

'I must go to the land of I-Know-Not-Where and bring back I-Know-Not-What,' he explained.

'Alas, my son,' sighed the old witch. 'Such a place and thing are quite beyond my ken, and in all the wide world there is only one more wise than I. That is the old green frog who has dwelt in the Green Marsh for nigh on three hundred years. But go now to sleep, for the morning is wiser than evening.'

When Petrushka was sound asleep on the stove, Baba Yaga took a pair of broomsticks and flew swiftly through the air to the great Green Marsh, and standing on the edge, she called out in a loud, shrill voice:

> 'Babushka
> Lyagushka
> Skakushka!*
> Art thou still among the living?'

The dark mud stirred and a deep croak emerged like a roll of distant thunder.

* i.e. 'Old Mother Hopping Frog.'

'I am.'

'Dost thou know the place I-Know-Not-Where?'

'I do.'

'And dost thou know the thing I-Know-Not-What?'

'I do.'

'Then wilt thou take my son-in-law to I-Know-Not-Where and help him to fetch back I-Know-Not-What?'

'I am old, Baba Yaga,' croaked the frog, 'and three hundred long years have weakened my legs. I can no longer leap from here to I-Know-Not-Where, but if thy son-in-law will carry me in a jug of fresh milk to the red River of Fire, I shall lead him thence.'

Baba Yaga took Babushka-Lyagushka-Skakushka and flew swiftly home on her broomsticks. There she prepared a jugful of fresh milk, placed the old frog in it, and woke the Archer.

'Take this jug of fresh milk with the wise old frog in it,' she said. 'Mount my horse and when she has brought you to the River of Fire, take Babushka-Lyagushka-Skakushka from the jug and she will tell you what to do.'

Holding the jug carefully in his hands, Petrushka mounted Baba Yaga's horse. Did they gallop a long way or a short way? We do not know, but eventually the clouds in the sky became tinged with red and the air grew hot and the Archer saw that the ground was littered with the bodies of birds with scorched wings. Suddenly, turning a corner, he saw before him the blazing, raging torrent of the red River of Fire.

The Archer shielded his eyes from the glare and dismounted from his horse.

'Let me out now!' croaked the frog.

The Archer took Babushka-Lyagushka-Skakushka from the jug of milk and placed her on the ground, whereupon the frog began to swell and swell until it was the size of a horse.

'Mount on my back, Archer!' she said.

Petrushka climbed on to the frog's back, and it continued to swell and swell until it was the size of a rick of hay.

'Art thou holding on tight, Archer?' croaked the frog.
'I am.'

The frog continued to swell and swell until it was as tall and broad as a forest of tall trees, and then, with a thud that made the whole wide earth tremble, it kicked down hard with its long legs and soared high into the sky above the River of Fire. Petrushka clung on for dear life and felt the flames scorching the soles of his feet, but the frog sailed safely across the blazing river and landed at a goodly distance on the other side. Here the frog breathed slowly out and out until it regained its normal size, and Petrushka stood once more upon his own legs.

'This is I-Know-Not-Where,' said Babushka-Lyagushka-Skakushka. 'Thou must walk on until thou comest to a building that is neither palace, house, nor hut. Enter in and hide behind the stove, and there thou shalt behold that for which thou has come. For there dwells I-Know-Not-What!'

Petrushka left the old frog and walked on and on past palaces, mansions, and lowly huts until he came to a tiny hovel made entirely out of twigs. There was no door in the doorway nor glass in the windows, and the Archer walked boldly in and hid behind the stove. A short time passed. Suddenly the whole hut shook and through the doorway walked a fierce old man with a long white beard. The Archer crouched down low behind the stove.

'Hey! Nobody!' cried the old man. 'Bring me food!'

And straightway in answer to his strange summons plates and dishes appeared on the table as though from nowhere, and in less time than it takes to tell, the table groaned beneath a meal fit for a king.

'Hey! Nobody!' called the old man again. 'Bring me drink!'

This time a large mug of beer appeared on the table, whereupon the old man finished his meal, dried his whiskers, and went out again into the forest.

The Archer emerged cautiously from behind the stove.

'Hey, Nobody!' he cried. 'Bring me food!'

Once again the plates and dishes floated through the air on invisible hands, the meat was served and bowls of fruit laid on the table. Petrushka sat down to eat.

'Hey, Nobody!' he cried, delighted at his success. 'Give me to drink!'

Again the invisible hands went to work and placed a large beaker of cool beer in front of the Archer. When the latter had finished his meal and the plates and dishes disappeared into thin air, Petrushka called out to Nobody to accompany him and then walked out of the hut together with his mysterious new servant.

He had not walked far when he came to a thick wood, and there, leaning against a tree, stood a huge bandit. A massive club rested on his shoulder.

'Halt!' cried the bandit in a threatening voice. 'Stop and give a wayfaring man a bite to eat and a drop to drink!'

'Hey, Nobody!' cried the Archer. 'Bring the wayfarer food and drink!'

No sooner said than done. A meal fit for a king spread itself in front of the bandit. There was borshch soup, fine veal cooked in cream, choice pirozhkis to eat, and mead and wine and beer to drink, and the bandit ate the meal greedily.

'That's a useful servant you have there,' he said, licking his lips and belching contentedly. 'I'll give you my club for your servant.'

'What is special about your club?' asked the Archer.

'There is not another like it in the whole wide world. You have only to say: "Club, kill such and such a man!" and the club will fly through the air, and beat the fellow to death, and then return to you.'

'A useful club indeed!' said Petrushka, and giving Nobody to the bandit in exchange for his magic club, he went off into the wood. Then he stopped.

'Club!' he said. 'Kill that bandit!'

The massive club whirled through the air, hit its late master a mighty blow on the temple, and struck him stone dead on the spot. Then it returned and fell at the Archer's feet.

'Hey, Nobody!' called out the latter. 'Return to me!'

And straightway his servant returned.

The Archer had not gone far into the dark wood, however, when another bandit suddenly stepped out of the bushes. A gusly was slung over his shoulder.

'Halt!' cried the bandit. 'Give a poor wayfaring man a bite to eat and a drop to drink!'

Once again a meal fit for a king spread itself before the bandit, and when he had eaten it greedily, he watched curiously as the plates and dishes and beakers disappeared into thin air.

'That's a useful servant you have there,' he said. 'Change him for my gusly!'

'What is special about your gusly?' asked the Archer.

'This is no ordinary musical instrument. If you pluck the first string, a great blue sea will appear before you; pluck the second string, and a fleet of ships will appear on the sea; pluck the third string, and the ships' guns will fire wherever you command.'

'That is indeed a useful instrument!' said the Archer, and giving Nobody to the bandit in exchange for the gusly, he went off into the forest. Then he stopped.

'Club!' he cried. 'Kill that bandit!'

No sooner said than done. The club whirled again through the air, struck the bandit a mighty blow on the temple, and he fell down dead.

'Hey, Nobody!' called Petrushka. 'Return to me!'

And thus it was that the Archer became the master of the magic club, the magic gusly, and the incomparable servant Nobody. For many days and nights he journeyed on, and eventually he returned to his native land. One can well imagine how glad he was to see his tiny hut again after such a long journey and how he looked forward to his fair wife's welcome, but alas! when he knocked on the door of his hut, there was no answer. He knocked again, but all was still inside. When he tried the door, however, he found that it was open after all, and he walked in.

Alas! here was no warm fire to greet him, and no fond

wife to welcome him home. The room was in great disorder, tables and chairs were overturned and the cups and dishes smashed to smithereens on the floor, and, worst of all, of the fair Dove Maiden there was no sign whatsoever. The Archer sat down in the midst of the ruins and wept aloud.

'Alas! Alas!' he cried. 'The ignoble Tsar has stolen my wife!'

As he said this, however, he felt something alight on his shoulder. It was a white dove! It kissed him first on his lips and then, before his wondering eyes, it changed into his beautiful wife Masha. The young couple embraced and wept for joy, and when Nobody had put everything in the little hut in order, Masha told her husband all that had happened.

'The Tsar's counsellor is in league with a wicked sorcerer who frequents the royal taverns, and on his evil advice he told the Tsar to profit by your absence on your supposedly fruitless quest and take me by force. But when they arrived to take me, I changed back into a dove and flew away.'

'Ach, the wicked Tsar!' cried the Archer. 'I never want to set eyes on him again!'

And commanding Nobody to build them a fine abode, he saw a superb palace of white stone rise from the ground, surrounded by a beautiful rose garden with a silver lake and a white swan. And there the Archer and the Dove Maiden lived in peace with their faithful servant Nobody.

It was not long, however, before the Tsar, one day while hunting, chanced to pass by the large white palace of white stone.

'Chort vozmí! The Devil take it!' he cried. 'Who built that on my land?'

'The palace belongs to your Archer, Your Majesty,' said an old peasant who was cutting wood in the nearby forest. 'The one who went to I-Know-Not-Where and brought back I-Know-Not-What.'

'Does it indeed!' snorted the angry Tsar. 'Then we'll soon have him out of there!'

But when he sent ten of his men to drag the Archer and

his wife out of their palace, the magic club came spinning out of the window and knocked them down like ninepins.

'Oho!' said the Tsar. 'The Archer means to fight, does he? Call out the Army!'

When the Archer saw the troops of the Tsar marching on his palace and preparing to shoot their arrows, he picked up his magic gusly and plucked the first string. Immediately a vast expanse of blue sea spread over the land between him and the Tsar's army. The Archer plucked the second string, and a fleet of stout wooden ships ranged themselves opposite the Tsar. The Archer plucked the third string, and with a deafening roar the guns of all the ships fired ball after ball into the ranks of the Army. Soon the field was heaped with the corpses of the dead, and when the remnants had fled for their lives, only the terrified Tsar and his wicked counsellor remained on the battlefield, forlorn as two lost souls in Hell.

'Archer!' pleaded the Tsar. 'Spare me! I meant no wrong! I will do anything you say!'

'Put your evil counsellor to death then, O Tsar!' cried the Archer, 'for he has sorely misled you and given you the Devil's counsel.'

'It shall be done!' cried the Tsar, but the evil counsellor, seeing that he had fallen from the Tsar's favour, took to his heels and ran. The Archer, however, called out to his magic club, and it whirled through the air, struck the wicked man a mighty blow on the temple, and killed him on the spot.

The Archer closed the gusly, and the ships and the vast blue sea disappeared. The Tsar was now a wiser man, and the Archer invited him into his palace, where the Tsar apologized for his misguided deeds. The Archer showed him I-Know-Not-What from I-Know-Not-Where and all that Nobody could do. A great banquet was prepared to which all the princes and nobles of the realm were invited; the ox roasted merrily on the spit and the green wine flowed abundantly, and all that could be heard above the merry laughter were the many calls of 'Nobody, here!' and 'Nobody, there!'

But if you ask me where Nobody really came from, I can only say I-Know-Not-Where, and if you ask me what manner of a creature he was, I can only say I-Know-Not-What. Perhaps the wise old frog in the Green Marsh could have told us more, but she has been dead now many a year.

Sorrow

IN a little village in the neighbourhood of Smolensk there
once lived two brothers. Each of them had been left an
equally large sum by their father on his death, but
through no great fault or merit of their own, the affairs of the
one declined while the affairs of the other prospered, and by
the time our story begins, the first brother had barely
enough to keep body and soul together, whereas the second
brother was rich and lived off the fat of the land. The poor
brother, however, although he had not so much as a crust of
bread in the whole of his lowly house, was rich in one thing
—children. They would swarm over the house and squawk
and bawl all day long, and though the poor muzhik made
valiant efforts to satisfy their constant hunger, they began to
drive his good wife and himself to distraction. One day,
when his children were crying and asking for food, the poor
man decided to swallow his pride and go and ask his rich
brother for help, and so, putting on his old fur cap and
much-worn coat, he went up to his brother's house and
knocked on the door.

'Brother,' he said, when the rich peasant opened the door.
'I am ashamed to beg from my own brother, but the truth
of the matter is, I have not the smallest morsel of food in the
house and my wife and our poor children are nigh on starving.'

'Brother,' replied the other. 'I would very much like to help you, but my affairs have been going very badly lately and I have not a copeck to spare.'

He paused and scratched his nose and looked out of the corner of his eye to see if his brother guessed he was lying. But his brother stood on the threshold fidgeting with his cap, his sunken eyes downcast in shame. A pang of pity almost penetrated the rich man's heart, but money, after all, was money!

'I tell you what, brother,' he said. 'Tomorrow is my name day, and though the Lord knows I can ill afford it, I have arranged a feast for my family and friends. Bring your wife along and have a good time.'

'Brother, how can I come to your feast?' objected the poor man. 'Your guests will be rich merchants in valuable fur coats and leather boots, whereas my wife and I have nothing but our rags and bast shoes.'

'That does not matter a bit!' said the other, determined to be generous now that it would be obvious to everybody. 'You are my brother and you shall come to my feast.'

The poor man returned home.

'I don't believe it!' cried his wife, when he told her she must prepare her best clothes because they had been invited to a banquet. 'Who would invite us?'

'My brother has done so. Tomorrow is his name day.'

The next day they made their way to the rich brother's house, offered their congratulations, and sat down at the table. Many well-to-do and famous men of the village were present, and to these the host paid great attention and treated them in a most princely fashion to food and drink. Somehow, however, he forgot all about his brother and his wife, and though the servants passed them by with large dishes of meat and poultry and fine goblets of wine which they deposited in front of the other guests, the poor couple said nothing, but sat there like a pair of frightened grey mice, staring at the empty space on the table before them. Eventually the banquet came to an end, and when the other guests began to take their leave of the rich peasant and to thank him for the fine repast, the poor brother and his wife rose also, and bowing low to their rich brother, followed the merry throng out into the snow. Their fellow guests, richly wined and dined, struck up a happy song, while the poor man and his wife followed after with empty bellies.

'Come on, wife!' said the poor man. 'Let us pretend we're well-fed and merry, and let us too sing a song!'

And when he opened his mouth and sang, two voices blended as one.

'You sing well, wife!' said the muzhik.

'I am not singing at all,' replied his wife, sadly. 'I haven't the heart.'

'Not singing?' said Misha, dumbfounded. 'Then who was?'

'I heard no one.'

'*Someone* was singing with me,' insisted her husband. 'I heard him.'

'The sight of so much wine must have made you drunk!' laughed his wife. 'But sing again, and I will listen.'

Misha opened his mouth and sang in his deep, rich voice, and once again two voices were heard—one deep and rich and one high and shrill. And both the peasant and his wife heard it.

'Sorrow!' cried the poor man. 'Is it you who sing with me?'

'Yes, Misha, it is I, Sorrow,' replied the shrill voice. 'And I am staying with you for ever!'

So Misha and his wife returned home with Sorrow, and no sooner were they back in their little hut than their un-invited guest asked the peasant to take him to the tavern that they might drink some vodka together.

'How can I?' said Misha gloomily. 'I have no money for vodka.'

'Sell your old fur coat, man!' said Sorrow. 'Another six months and it will be summer, and you won't need it any more.'

So Misha and Sorrow went together to the tavern and drank away the peasant's old fur coat.

The next day Sorrow again asked the peasant to go to the tavern and drink vodka with him.

'How can I?' said Misha. 'I have no money for vodka.'

'Sell your sledge,' said Sorrow. 'Another six months and it will be summer, and what should you do with a sledge when there is no snow on the ground?'

So Misha and Sorrow went together to the tavern and drank away the peasant's old wooden sledge.

And so it went on. Every day Sorrow took the peasant away to the tavern, and before the month was out, Misha had sold all he possessed—his plough, his harrow, his cart, his wife's rings, and even his old fur cap—for a few kegs of vodka.

The next time that Sorrow proposed to go to the tavern, therefore, Misha said:

'This time it is really impossible, Sorrow, for I have no money whatsoever and not a single thing to sell to get any.'

'Don't worry about that,' replied Sorrow. 'Go and ask your neighbour to lend you his ox and cart.'

Misha did as Sorrow bade him, and when he returned with his neighbour's ox and cart, Sorrow mounted beside him and together they drove out into the open countryside.

They had not gone very far when Sorrow plucked the peasant's sleeve.

'Do you see that large rock in the middle of that field?' he said. 'Drive up there.'

Misha whipped the ox and drove across the field up to the rock.

'Now get down,' said Sorrow, 'and help me lift up the stone.'

Misha descended from the cart, and between them the two managed, though with great difficulty, to roll the huge rock to one side, and there where the rock had lain was a large pit, and in the pit lay a mass of golden coins.

The peasant's eyes almost popped out of his head to see such wealth.

'Well, let's not waste time looking!' said his companion. 'Start loading the cart!'

Misha got to work with a shovel, and before long he had emptied the pit of all the golden treasure.

'Sorrow,' he said finally, mopping his brow. 'See if there is any more gold left down there, there's a good fellow. My back is killing me.'

'I can't see any,' replied Sorrow.

'Isn't there something shining behind those pebbles?' said the peasant. 'Go down and see what it is.'

Sorrow grumbled a little, but thinking of the vodka represented by one gold ducat, he slid down into the hole in the ground. And no sooner had he done so than Misha hurled himself against the huge rock and rolled it back on top of the pit.

'Hey, let me out!' shouted Sorrow. 'What's the idea?'

'You stay there, Sorrow,' said the peasant. 'You are much too expensive a guest.'

And chuckling to himself, Misha drove off home, tipped the golden treasure into the cellar, and returned the ox and cart to his neighbour. The next day he bought up great possessions, and within a short time he was living in great luxury in a fine mansion.

When the rich brother got to hear of the sudden change in his brother's fortunes, he did not delay to pay him a visit.

'Gospodi!' he said, when he spied the tall white towers of his brother's new abode and saw the rich tapestries and ornaments within. 'You are ten times as rich as I, brother! Tell me, how did you acquire such a vast fortune?'

Misha told him everything that had happened since the day he returned from the name-day celebrations.

'But now Sorrow is safe and sound in the deep hole in the ground,' he concluded, 'and as long as he stays there, I have nothing to fear.'

By now his brother was green with envy.

'Well, well,' he thought to himself, as he passed by the field where the great rock stood. 'It is not pleasant to have my brother lording it over me, and if I let Sorrow out of his prison, he'll soon rid Misha of his fortune!'

So walking over to the rock, he pushed and pushed until

the sweat ran down his face, and the rock rolled aside. No
sooner had it done so, however, than Sorrow hopped angrily
out of the hole, jumped on the rich peasant's shoulders and
clung tightly to his hair.

'Ah! Got you!' he cried. 'So you intended to let me
perish in that pit, did you? But you will not escape me
again, my fine fellow, that I can tell you!'

'Get off my neck!' cried the rich brother. 'I didn't shut
you in the pit. It was my brother!'

'What a liar!' cried Sorrow. 'You have deceived me once,
but you shall never get the better of me again!'

And so the rich merchant had to return home with
Sorrow seated firmly round his neck. The months went by,
the vodka flowed freely, and the roubles and kopecks
dwindled away, and the once-rich peasant realized that if he
did not soon rid himself of his obtrusive guest, he would
soon be ruined entirely. He thought and thought and wondered and wondered until at last, going into the yard, he
took two wooden bungs, drove one of these into one side of
the hub of a large cartwheel, and returned to the house.

'Sorrow,' he cried. 'Why do you lie on the stove all day
long?'

'What else is there to do?' yawned his guest. 'The tavern
is not yet open.'

'Come into the yard and let us play hide-and-seek together.'

'That's an excellent idea!' said Sorrow delightedly, and in
a flash he was down from the stove and in the yard. 'And I
was beginning to think you didn't like my company!'

First the peasant hid himself behind a barrel, but Sorrow
found him easily.

'Now it is your turn to hide,' said the peasant.

'What is the good of my hiding?' said Sorrow. 'You'll
never find me, for I can hide in the smallest nook and
cranny.'

'I don't believe it!' said the other, shaking his head. 'Why,
you couldn't even get into the hub of this cartwheel!'

'Couldn't I indeed?' said Sorrow, and in less time than it

takes to tell, he was crouching down in the place indicated, whereupon the peasant picked up the second bung and hammered it swiftly into the hub of the wheel.

'Hey, let me out!' cried Sorrow. 'I can't move an inch!'

'You shall never get out of this,' said the peasant grimly, and lifting up the wheel, he rolled it off to the river, pushed it into the water, and watched it float away with the swift stream.

Was Sorrow drowned? Well, no one can say for sure, but although it seems unlikely that he disappeared for ever from Russia, there is no account of how he escaped from the wheel. In any case, however, it is established that he never returned to trouble the two brothers in the little village near Smolensk, for the one continued to enjoy his newly-found wealth, while the other worked hard to repair the misfortune that envy and Sorrow had brought upon his affairs.

The Mountain of Gold

THERE was once a young muzhik named Grishka who, fallen on evil days, slung a shovel across his shoulder and marched out into the world to look for work. When he arrived at the first largish town, he made his way to the market-place and stood there in the hope that someone would hire his labour. He was not alone in his poverty, however, and with him on the market-place stood a throng of young fellows like himself who also waited for work. An occasional merchant would come and look them over and hire one of their number, but although Grishka waited patiently for some hours, he was not chosen, and he began to give up hope. Suddenly, however, a rumbling of wheels was heard on the road and a richly-attired merchant drove up in a solid gold coach drawn by two splendid white horses. Expecting all his fellow labourers to rush towards the merchant and beg for his hire, Grishka stepped back cautiously, but it was with no little surprise that he saw all his companions take to their heels and make off down the side streets as fast as their legs would carry them, and in less time than it takes to tell, Grishka found himself all alone on the deserted market-place.

The magnificent carriage drew to a halt in front of him and the merchant poked his head out of the window.

'Hey there, young fellow!' he cried. 'Are you looking for work?'

'That was my intention,' replied the lad.

'How much would you want to work for me?'

Grishka thought a little.

'One hundred roubles a day,' he replied at last.

'One hundred roubles a day!' exclaimed the merchant. 'Why is your hire so expensive?'

'If you can get anyone else to work for you for less, you are free to do so,' replied Grishka. 'But I noticed that everyone fled at your approach.'

'All right, all right!' snapped the merchant testily. 'A hundred roubles a day it is, then! Meet me at the harbour at seven sharp tomorrow morning. I shall be waiting for you.'

And with that the merchant drew back his head and the carriage drove rapidly away.

The next morning Grishka went to the harbour at the time arranged, boarded the merchant's ship, and sailed with him out to sea. It is not known how long they sailed, but eventually the shape of a rocky island appeared on the horizon, and as they drew nearer, Grishka saw something burning like fire on the shore.

'What are those flames?' he asked.

'They are not flames,' replied the merchant. 'That is the reflection of my golden palace.'

When they had anchored and disembarked on the island, the merchant's wife and daughter were there on the quay to meet them, and Grishka felt a wave of happiness flow through him, for the maiden was of a beauty far beyond the imagination or skill of any storyteller. When they had greeted each other, they went up together to the golden palace and sat down to eat, drink, and make merry.

'Drink, Grishka,' said the merchant. 'Today we feast, but tomorrow we work.'

Now as one may already have guessed, Grishka was a very

handsome youth, tall and strong and healthy, and if he himself had been struck by the beauty of the merchant's daughter, she was by no means indifferently disposed towards him. On the contrary, she fell head over heels in love with him at first sight, and he with her, and when the merchant began to nod off after the heavy meal, the girl led Grishka into the adjoining room and pressed a tinderbox into his hand.

'What is this for?' asked the youth.

'Take it with you wherever you go,' whispered the merchant's daughter. 'If you ever get into any trouble on this island, strike the steel against the flint and help will come immediately.'

The following morning the merchant took Grishka to the great Mountain of Gold, and seating himself at its foot, bade him climb to the summit, but try as he might, Grishka slipped and slithered and could make no headway whatsoever.

'Yesterday's banquet was too much for you, young man, that's clear!' said the merchant. 'Here, have a drink to restore your strength.'

Grishka accepted the proffered glass, but as soon as his lips touched the brim, the glass dropped from his fingers and he fell into a deep slumber. Hereupon the merchant took out his knife, slit the throat of the sorry nag that had accompanied them thither, removed its entrails, sewed his labourer's unconscious body inside the horse's belly, and quickly hid himself behind a nearby bush.

After a while, two enormous black crows alighted on the horse's carcase, and seizing it in their iron beaks, they bore it away to the top of the mountain where, settling down to feed, they tore away at the horse's flesh until the bones were bare. Then Grishka awoke, drove the crows away and looked about him in amazement.

'Where are you, merchant?' he cried.

'Here I am, Grishka,' shouted up the merchant at the foot of the mountain.

'And where am I?'

'You are on top of the Mountain of God,' came the reply. 'Take your spade and start digging.'

Grishka took his spade and dug vigorously, rolling the gold down the mountain-side. The merchant loaded it on to the carts, and by evening they were all full to the top.

'That's enough, Grishka!' shouted the merchant. 'You can stop now. Thanks for your help.'

'Don't mention it,' said Grishka. 'But what am I to do now?'

'Just stay up there and wait, my lad!' laughed the merchant evilly. 'Ninety-nine fine fellows have already perished up there, and you'll make it a hundred!'

And chuckling to himself, the wicked merchant went off with his gold-laden carts and spared not a thought for Grishka's sorry plight.

'This is a fine thing!' reflected the young man, watching the two great crows circle above him as they waited for him to grow too weak to resist their powerful beaks. 'If I cannot get off this mountain, I shall surely die of hunger.'

Suddenly, however, he remembered the gift the merchant's daughter had given him, and taking the tinderbox from his pocket, he struck the flint against the steel. Before the sparks had died away, two shining youths appeared before him.

'What is thy will, master?' they asked.

'Take me down from this mountain,' said Grishka. No sooner was it said than done, and Grishka found himself on he seashore below. A ship happened to be sailing past the island, and Grishka waved frantically to attract the crew's attention.

'Hey there, good sailors!' he cried. 'Take me with you!'

'Sorry, brother!' replied the sailors. 'We can sail a hundred versts in the time it takes to pick you up, and we have not a minute to spare!'

As the ship made to sail past the island and leave him behind, however, Grishka struck his magic tinderbox: a powerful wind blew up, a dark storm threatened, and the ship tossed about like a cork on the churning surface of the sea, quite unable to make any headway.

'We shall have to go back and take that man aboard,' said the sailors. 'He is clearly no ordinary human being.'

And returning to the island, they took Grishka aboard and sailed with him to the town whence he came. When some time had elapsed—much or little, we do not know—Grishka again shouldered his spade and stood in the market-place looking for work as before. And again the rich merchant drove up in his fine golden carriage, and again the crowd scattered and left Grishka alone on the market-place.

'Are you looking for work, young fellow?' said the merchant.

'That was my intention,' replied Grishka.

'How much would you want to work for me?'

'Two hundred roubles a day.'

'Two hundred roubles a day! But that is twice as much as I've paid anybody!'

'If you can get anyone else to work for you for less you are free to do so,' replied Grishka. 'Though I noticed that everyone fled at your approach.'

'All right,' grumbled the merchant. The price did not matter much in any case. 'Meet me at the harbour at seven sharp tomorrow. I shall be waiting for you.'

So once again Grishka sailed to the island, accompanied the merchant to the foot of the Mountain of Gold, and was offered a sleeping draught 'to restore his strength'. But this time Grishka knew better.

'You drink first, master,' he said, and poured the merchant a sleeping draught from his own water-bottle.

Suspecting nothing, the merchant took the glass, and as soon as his lips touched the brim, he dropped off into a deep sleep. Then Grishka took his knife, slit the throat of the horse that had accompanied them, removed its entrails, sewed up the merchant in the horse's belly, and hid behind a nearby bush.

The enormous crows appeared, and this time it was the

merchant who awoke to find himself in a cage of bare bones on top of the mountain.

'Where am I?' he cried.

'On top of the Mountain of Gold,' shouted Grishka. 'And now, goodbye!'

'Don't leave me!' cried the merchant. 'I shall die!'

'Ninety-nine fine fellows have perished up there,' mocked Grishka. 'But you, not I, shall be the hundredth!'

And returning to the golden palace, Grishka took the merchant's beautiful daughter to be his wife. The wicked merchant, however, remained on the bare mountain and watched the black crows circling above him, drawing closer and closer as he grew weaker and weaker with hunger, until at last they fell upon him and tore him to shreds with their sharp beaks. Which, after all, was only just!

The Russian and the Tatar

A RUSSIAN and a Tatar were travelling through the
Southern steppes when night fell and obliged them
to camp out in the open air. Dismounting from their
horses, they put up their tent, made a fire, and ate and talked
together until they grew sleepy.

'Who is going to keep watch over the horses?' yawned the
Tatar. 'Let us draw lots for it.'

'Why should I keep watch over the horses?' said the
Russian. 'My horse is white, and if anything happened, I
should easily see it in the darkness. Yours, however, is
black, so you had better keep watch.'

'Your horse is as thin as a rake and mine is a good sturdy
beast,' replied the Tatar. 'But I am so tired that I would
willingly exchange mine for yours.'

'Done!' said the Russian, shaking hands to clinch the
bargain. 'But there is still no need for me to keep watch.
If a thief should come, he'll never find my black horse in the
darkness, but he'll soon find yours all right.'

The Tatar was still trying to work out how he had been
tricked even after the Russian had been snoring loudly for
some time.

'I'll get my own back on this Russian rogue!' muttered
the Tatar, and rising softly, he led the Russian's horse

away, drove it into the swamp, returned, and fell soundly asleep.

The Russian, however, had only pretended to be asleep, and had in fact observed all that the Tatar had done. When the latter was snoring loudly and stirred no more, the Russian rose, retrieved his own horse from the swamp, left that of the Tatar in its place, returned, and went to sleep.

In the middle of the night the Tatar awoke, grinned broadly, and shook his companion by the shoulder.

'Wake up, brother!' he said. 'I have just had a terrible dream. I dreamt that a devil came into our camp and drove your horse into the swamp.'

'That is strange, brother,' replied the Russian. 'I too have just had a dream. I dreamt that *your* horse broke loose and wandered into the swamp and drowned.'

The Tatar leapt anxiously to his feet and rushed to the swamp. It was too late. His horse had perished!

When dawn came, the Russian sold his horse so as to keep the Tatar company, and they resumed their journey on foot.

'I have an idea,' said the Russian after a while. 'Let us take turns in carrying each other on our backs.'

'But how shall we measure the distance?' asked the Tatar.

'Let the one who is carried sing a song until he drops off to sleep,' said the Russian, 'and then it shall be the turn of the other.'

They threw lots, the Tatar won, mounted on to the Russian's shoulders, and began to sing a lullaby.

'*Taldi-baldi, taldi-baldi, taldi-baldi,*' he sang, and before they had gone half a verst, his head bobbed and nodded, and he fell fast asleep. The Russian halted and slid him to the ground.

'My turn now, brother,' he said, and mounting on to the Tatar's shoulder, he began to sing a song.

'*Tili-tili, tili-tili, tili-tili,*' he sang, and was still singing when the Tatar had carried him for a distance of twenty versts.

'How long does this *tili-tili* song of yours go on for?' asked the Tatar

'Oh, at the most another ten versts,' replied the Russian. 'It's funny, but somehow I don't seem to get tired.'

The *tili-tili* song was still proceeding when night fell again and it was time to camp. When the fire was lit and they came to prepare the meal, however, it was discovered that there was nothing left except one small chicken which was clearly not big enough to satisfy two grown men.

'Listen, Russian!' said the Tatar. 'Let us go to sleep for an hour and he who dreams the best dream shall have the chicken all to himself.'

'That's a good idea,' agreed the Russian, and they both lay down to pretend to sleep. The Tatar thought up a good dream, but while his back was turned, the Russian ate up the chicken by himself and went contentedly to sleep.

When the hour was up, the Tatar roused his companion.

'Wake up, brother!' he said. 'Let us tell each other what we have dreamed.'

'Tell yours first,' said the Russian.

'I had a marvellous dream,' said the Tatar. 'I dreamt I was floating through the heavens at night among the host of stars; two shining angels held up my feet and four fair maidens led me by the hand into a wonderful palace in the middle of a sweet-smelling garden of flowers. How beautiful it all was!'

'Yes,' said the Russian. 'I had the same dream, and naturally thinking you would never return to earth again, I ate the chicken.'

The Tatar packed his bag, left the Russian, and henceforth travelled alone.

The Soldier Who Did Not Wash

THERE was once an old soldier who, after serving the Tsar faithfully for many long years, was finally released and sent on his way without so much as the price of a sucked egg. And as he journeyed disconsolately homewards, his legs felt tired and his feet grew sore and he stopped and sank down wearily beside a lake.

'Well,' he reflected gloomily, 'what am I to do now? I have no money, no food, no roof over my head, and am still many miles from home. At such a time a man feels like hiring himself out to the Devil!'

No sooner had he uttered these words, than a tiny, hairy, and incredibly ugly devil popped out of the lake and sprang ashore.

'Greetings, Soldier!' said the newcomer.

'What do *you* want?' asked the Soldier.

'Did you not say you were thinking of hiring yourself to the Devil? We'll pay you well if you do, you know!'

The Soldier scratched the back of his head and thought the matter over.

'What sort of things would I have to do for you?' he asked cautiously.

'Do? You won't have to *do* anything—quite the contrary in fact. All that is required is that you undertake not to wash, shave, comb your hair, cut your nails, wipe your nose, or change your clothes for the space of fifteen years. And in exchange for this we will provide you with anything you ask of us for the same term.'

'Hm!' murmured the Soldier. 'What you suggest is a small boy's paradise, but it does not come easy to an old soldier like me. It is not what they taught us in the barracks. However, what goes into the head can come out again, so I accept your kind offer. But no treachery, mind!'

'Have no care for that, Soldier! We shall not fail to keep *our* side of the bargain. If *you* fail, however, the Devil shall keep your soul for all eternity.'

'Done!' cried the Soldier. 'Now show willing by fetching me a good pile of roubles and get me off to Moscow as quickly as you can!'

The tiny devil plunged back into the lake and emerged with a huge sack of silver. Then, bearing the Soldier swiftly off to Moscow, he put him down in the main square, and disappeared into thin air.

'Ho-ho! There's a fool for you!' chuckled the Soldier. 'I've done no work whatsoever, yet here I am in Moscow, with plenty of money!'

The Soldier hired a richly appointed mansion to live in all alone, and straightway, in accordance with his bargain, ceased to wash, shave, comb his hair, cut his nails, wipe his nose, and change his clothes from that day forth. As time went on, of course, his appearance grew more and more hideous, but so well did the Devil keep his side of the bargain that before long the Soldier was living the life of a Grand Duke, and grew so wealthy that he soon had no room to store all the gold and silver he amassed at the Devil's expense.

'I'll help the poor,' decided the Soldier, 'and they will pray for my soul!'

So the Soldier began to distribute his wealth among the poor, but though he handed it out generously to all the beggars, widows, and orphans he could find, his wealth did not diminish, and his fame spread throughout the whole land of Russia. For fourteen years the Soldier lived thus, and when the fifteenth and last year of the agreed term of service arrived, it so happened that the Tsar's Treasury ran short of funds. The Tsar, therefore, knowing of the Soldier's great wealth and choosing in the circumstances to ignore his eccentricities, summoned him to appear at court. Unwashed, unshaven and unkempt, with running nose and filthy clothes, the Soldier came before the Tsar.

'Greetings, Your Majesty,' said the Soldier, bowing low.

The Tsar shrank back in his throne and but for the fear of offending a potential benefactor would dearly have loved to hold his handkerchief to his nose.

'Soldier,' said the Tsar, clearing his throat. 'Everybody tells me how good you are to the people, and alas! it so happens that their Tsar has no money wherewith to pay the Army. If you could lend me some, I should make you a general.'

'Your Majesty will forgive me if I say that I have not the slightest desire to join the Army again, be it as a general or anything else!' replied the Soldier. 'However, if you really wish to reward me, marry me to one of your daughters. Then you may take as much of my money as you desire.'

The Tsar was sorry for his daughters, but there was nothing for it: he had no money.

'So be it!' he said, mopping his brow. 'Have a portrait of yourself made so that I may show it to the three princesses. Then we shall see which one shall choose to be your bride. And, Soldier, if I were you I'd get a really *good* portrait done. I can give you the address of an artist who has done many portraits for the Court. He's a bit short-sighted, but he has an excellent imagination!'

When the Soldier withdrew from the royal presence, he did not go to the artist the Tsar recommended, but ordered a really spitting image to be made which concealed nothing of his filth and squalor. When the Tsar took it and showed it to the eldest daughter, she took one look at the horrible creature with matted hair and running nose, and shuddered in disgust.

'Marry *him*?' she cried. 'I'd marry the Devil first!'

And no sooner had she said these words than the tiny devil popped up behind her and noted down her soul in his little black book.

Nor was the second princess any more anxious to marry the Soldier, and rashly expressing a like preference for the Prince of Darkness, she too had her name taken down by the tiny devil.

When the Tsar showed the portrait to the third princess, however, she—for in fairy-tales the youngest daughter is always of the sweetest possible disposition and very obedient to her parents' wishes—she only sighed and looked unhappy.

'If such is my fate, Father,' she said, 'I shall marry this Soldier, and may God have mercy on me.'

The Tsar was overjoyed and sent word right away to the Soldier to prepare for the impending wedding. He also sent a dozen carts along with the messenger at the same time to serve as a tactful reminder of their bargain, and these the Soldier filled with devils' gold and sent them back to the palace. The honour of the Tsar was saved, and full of gratitude, he fell to inviting the Soldier every day to dinner, sat him at his own table, and ate and drank in his company.

And thus, while preparations for the royal marriage were in full swing, the Soldier's term of fifteen years' service to the Devil came to an end and he summoned the little demon who had actuated the contract to appear before him.

'Well, bogy!' he said. 'My fifteen years are up, and since I have no wish to renew the contract, I should be obliged if you would restore me to my original shape.'

The Devil could not persuade the Soldier of the multiple advantages of remaining in the service of the Prince of Hell, so with a reluctant sigh he took the Soldier, broke him into little pieces of flesh and bone, threw them all into a large cauldron and began to boil them up. When all the pieces were clean, he took them out of the pot, assembled all the parts in their proper order, bone to bone, joint to joint, and sinew to sinew, and sprinkled them with the waters of life and death. The Soldier rose from the ground as handsome a youth as ever was, and thanking the Devil for his services, he married the young princess and lived happily ever after.

And I was at their wedding
And mead and beer drank I.
Wine was also there to drink
And I drank the barrels dry.

When the Devil returned to the lake, however, he was ordered by his grandfather to give an account of the transaction, and when the old devil heard that the lease on the Soldier's life had expired and that the Soldier had benefited royally from the arrangement, he was furious.

'Are you telling me that in the course of fifteen years you couldn't find one way in which to trick this miserable human?' he bellowed. 'Just think of all the good red gold you've squandered! Hey there, fellow demons! Take this fool and fling him into the lake of boiling pitch!'

'Grandfather, wait!' cried the unfortunate Devil. 'The Soldier has only one soul, but I have got two in his stead!'

'How is that?'

'The Soldier wanted to marry one of the King's daughters, but two of them declared they would prefer to marry the Devil instead. And so these two nice, fat, female Russian souls are ours!'

'Good lad!' said the grandfather devil, his wrath suddenly placated. 'Let him go, comrades, for the boy knows his business, after all!'

The Crock of Gold

ONCE upon a time there lived in Russia a poor old fellow and his wife, and one winter when the snow lay thick on the ground and the fuel burned low in the stove, the old woman took sick and died. Outside the frost was bitter cold, but the old man put on his thin, torn coat and began to tramp from one hut to another to beg his neighbours help him dig a grave for his dead wife. But alas! all the neighbours knew how poor he was and quite incapable of recompensing their labours, and all refused to help him and shut the door in his face. The old man touched his cap and limped sadly off, and eventually, when all else failed, he made his way to the church to ask the village priest for help. Now this priest was a mercenary man and—though it grieves one to have to say it—devoid of any Christian conscience whatsoever, so that when he saw the old man standing before his door in the snow asking for help to bury his dead wife, his first thought was how much he might gain from the affair.

'Where is the money to pay for the burial?' he asked. 'I must be paid in advance.'

'It would be a sin to hide the truth from you, Your Reverence,' said the old man. 'I possess not a single copeck in the world. But if you would wait a little, I'll earn the money and pay you back with interest.'

This, however, was not the priest's way of doing business.

'If you have no money, you old fool,' he cried angrily, 'how dare you come to me, a priest? Be off at once, and bury your old woman yourself!'

The old man went sadly home, took a spade and an axe, and going to the churchyard, began to dig a hole with his own frail hands. He hacked feebly away at the frozen earth with the axe and then began to dig with the spade. His heart beat painfully in his breast and everything began to swim before his eyes, when suddenly his spade struck against something hard. When he had recovered his breath and looked to see what it was, he could scarce believe his eyes: he had unearthed a crock of yellow gold, full of rich ducats that shone like fire!

'Praise the Lord!' cried the old man. 'Now I shall have enough money wherewith to bury my old wife and give a feast in her memory!'

So he dug no more, threw down his tools, and carried home the crock of gold; and sending his daughter-in-law off to purchase all manner of good things for the feast, he took a golden ducat from the pot and hobbled back to the priest.

'What did I tell you, old fool?' cried the priest when he saw the old man standing again in the doorway. 'What have you slunk back here for? Did I not tell you not to come to me in the hope of getting something for nothing? Be off with you!'

And he made to shut the door in the old man's face.

'Don't be angry, Your Reverence!' said the old man. 'I have brought you some gold. Please bury my wife and I shall never forget your kindness.'

When he saw the ducat shining in the old man's palm, the priest could hardly believe his eyes. He snatched the coin, held it up to the light and turned it this way and that, bit it, and banged it on the doorstep. When finally he was assured that the coin was genuine, he could not do enough to flatter and cajole the old man in a clumsy attempt to make amends for his rudeness—not for the sake of the gold, mind! but out of repentance for his sins.

'Don't worry, grandfather,' he smiled, 'everything shall be done.'

The old man thanked the priest and went back home.

'What do you think of that?' said the priest to his wife. 'He's as poor as a mouse, and yet the old fool pays me a gold ducat to bury that old crone of his! I have buried many a famous person in my time, but no one ever gave me a whole ducat before!'

So the priest assembled all his retinue and gave the old woman a decent burial. After the funeral, the old man invited the priest along to the pominki, the feast in memory of the dead, and all the neighbours came and sat at his table, which was laden with all manner of fine food and drink. The priest sat down and ate for three people, grasping to the right and left of him and gobbling away so greedily you would have thought it was his last meal on earth; and it was not until all the other guests had eaten and drunk their fill and gone back to their homes that the priest himself rose from the table. The old man accompanied him to the door, and when the priest saw that they were alone, he took him by the arm and lowered his voice.

'Listen, friend!' he whispered. 'I know everything is not as it should be, but if you will make your confession to me, your soul shall be cleansed of sin. Tell me, how did you manage to arrange things so quickly? Yesterday you were a poor peasant, and today you are rich. Where did all these riches come from? Whom have you murdered? Whom have you robbed? Confess now, or you will be damned for ever!'

'Your Reverence, I have neither murdered nor robbed,' replied the old man. 'I found a crock of gold, that is all.'

And he told the priest everything that had happened.

The priest returned home consumed with envy, and henceforth could think of nothing but how to find a way of cheating the old peasant of his money. At last he hit upon a plan.

'Listen, matushka,' he said to his wife. 'We have a goat, have we not?'

'We have.'

'Good! Let us wait till nightfall and then we will arrange things properly.'

When it was dark, the priest dragged the goat into the shed, cut its throat, and skinned it, horns and all.

'Get a needle and thread, wife,' he chuckled, 'and sew this skin round my body so that it won't slip off.'

The priest was a little weed of a man and the goat was a big one, so it was not long before he was neatly sewn up inside the goat's skin. At midnight, creeping stealthily up to the old man's hut, he began to scratch and tap at the window-pane.

'Who's there?' cried the old man, jumping up in alarm.

'The Devil!' replied the priest.

'This place is holy!' shouted the old man, and began to cross himself frantically and chant his prayers at the top of his voice.

'Listen, old man,' said the priest. 'However much you cross yourself, you shall not escape me. Give me back my crock of gold you stole from the graveyard, or you shall bitterly regret it. I had pity on you at the time of your bereavement, and thinking you would take only enough to cover the expenses of the funeral, I showed you my crock of gold. But you were greedy and took everything, and now you must give it back!'

The old man looked through the window, and his horrified eyes fell upon the horns and the beard of the goatskin! That it was the Devil himself, he had no doubt!

'Well,' he thought, shaking in every limb. 'I have lived up to now without money, and no doubt I shall manage well enough without it in future.'

Sighing deeply, the old man took the pot of gold, and carrying it outside, threw it to the ground and bolted back inside his hut.

The priest hastened home, greedily clutching his booty under his arm.

'We are rich! We are rich, wife!' he rejoiced. 'Now get me out of the goatskin before anyone sees me.'

His wife took a sharp knife and began cutting the skin along the seam.

'Ow!' cried the priest. He clapped his hand to his belly, and his fingers came away dripping with blood. 'Cut it in another place, woman. It hurts there!'

His wife ripped it open in another place, and the priest gave another agonized scream as the blood again spurted forth. His wife tried again and again, but nothing they could do could remove the goatskin from the priest's body. They tried taking the crock of gold back to the old man, but the skin clung to the wicked priest as tightly as ever, and thus he remained to the day of his death. Some in the village said that God punished the priest, and others say the Devil, but no one knew for sure, for the truth of the matter is that the wicked man had offended both!

One-Eyed Likho*

THERE was once a blacksmith who had never known any manner of misfortune or ill luck in the whole of his life, and when people spoke to him of Evil, he only shrugged his shoulders and looked blank, for he had not the slightest idea of what they meant. One day, however, when he felt more puzzled than usual, for his ignorance was beginning to prey on his mind, he thought of remedying this state of affairs.

'Everyone says there is a thing called Evil in this world of which I know nothing,' he thought to himself. 'I shall go and find out what it is.'

So fortifying himself with a good draught of vodka, he sallied forth into the wide world to look for Evil. He had not gone far when he caught up with a tailor.

'Greetings, friend,' said the tailor.

'Greetings, brother,' said the blacksmith.

'Where are you going?' asked the tailor.

'Well, everybody tells me there is something called Evil in the world, and since I have never seen it for myself, I am setting out to look for it.'

'That's strange,' said the tailor. 'I too have always lived

* Russian for 'Evil.'

148

in peace and have never encountered Evil. Let us go and look for it together.'

The two men walked on and on until they came to a narrow path through a thick, dark forest, and when they had walked some way along this path, they came to a large hut standing in a clearing. Night had already fallen, and in view of the fact that they had nowhere else to go, they decided to knock at the hut and ask for shelter. When they knocked, however, there was no reply, and when they had knocked again in vain, they pushed open the door and looked in. There was no one there. All was bare and deserted inside, so they sat down and waited for the owner to return. It was not long before a tall, thin, crooked old woman with only one eye in her head came into the hut.

'Aha! I see I have guests!' she cackled. 'Good evening to you!'

'Good evening, babushka,' said the two travellers. 'We should like to spend the night in your hut.'

'And so you shall!' croaked One-Eyed Likho, for she it was. 'I am going to eat you for my supper!'

The two travellers, now petrified with fear, watched with wide, rolling eyes as the old witch gathered up some twigs and lit the stove. Then, before he had time to duck or scream, she caught hold of the tailor by his hair, cut his throat, and put him in the oven to bake.

The good smith was at his wits' end. It was all too clear that a like fate awaited him also, for when One-Eyed Likho had picked his comrade's bones clean, she began to eye him ominously.

'Babushka,' he said awkwardly, 'I am a blacksmith.'

'Indeed?' said Likho, suddenly interested. 'What can you forge?'

'Anything! Anything at all.'

'Could you make me another eye?'

'I could,' rejoined the smith, cautiously. 'But you would have to get me some rope to bind you with, or you would never remain still. You see, I must hammer the new eye in.'

Likho rummaged about until she found two ropes, one

thick and one thin, whereupon the smith trussed her up with
the thin rope first.

'Twist about, babushka, and see if the rope will hold,'
said the smith.

As soon as the witch turned around, the rope snapped like
cotton, so the smith took the other rope and tied her up with
that.

'Now twist about again,' he said, and this time, though
Likho twisted and turned with all her might, the rope held
fast. When he saw that the witch was quite unable to escape,
the wily smith seized an awl, poked it into her one eye and
hammered it swiftly in with the back of his axe. The blinded
witch screamed in agony and rage, and struggled so violently
that the stout rope snapped. Likho groped blindly around
the corners of the hut in an attempt to catch the smith, but
somehow he managed to keep out of her reach.

'You shall not escape me long, villain!' she screamed,
sitting on the threshold nursing her blind eye. 'You can't
get out of this hut without passing me!'

The smith trembled and shook, for now he saw to his cost
what Evil was. After a time, however, Likho's sheep returned
from pasture and these she drove into the hut to sleep. The
smith, who could not escape from the hut while Likho sat on
the threshold, was obliged to spend the night there also.

On the morrow, Likho rose to let the sheep out to graze,
and to ensure that the smith did not attempt to escape in
their midst, she felt the back of each sheep as it passed her.
When the smith saw this, he took off his sheepskin coat,
turned it inside out, pulled it well over him and crept up to
the witch on all fours. When the witch's gnarled hand
touched his back, she thought it was another sheep and
pushed him out along with the rest. Once outside, the smith
was so overjoyed that he could not restrain his exultation.

'Farewell, Likho!' he shouted. 'You have taught me what
Evil is all right, but now you are powerless to do me harm!'

'Wait, villain!' shouted Likho. 'You shall learn still
better what Evil is, for you have not got away from me yet!'

The smith hurried back along the narrow path through

the thick, dark forest, but before he emerged from its depths he spied a gold-handled axe sticking in a tree, and felt an irresistible urge to seize it. As soon as he did so, however, his hand clung to the axe as though grown to it, and tug and pull as he might, he could not break loose. Suddenly, to his horror, he heard the sound of cracking twigs in the forest, and knew that it was Likho following in hot pursuit.

'Got you, you villain!' she shouted. 'You shall not escape me now!'

The smith's heart sank, but fear did not paralyse his brain, and in desperation he tugged a knife out of his pocket, hacked his hand clean off at the wrist, and ran away as fast as his legs would carry him.

When he returned to his village, the people crowded about him to know what had happened.

'My friend the tailor has been baked in an oven and eaten up,' he said, 'and I myself have left my good right hand hanging on Likho's axe!'

'Do you know what Evil is now?' asked an old villager.

'Aye, that I do!' said the smith ruefully. 'And I know that it is easier to find than to escape!'

The Death Watch

EVERY day, in accordance with the wishes of his father
the village priest, who concerned himself greatly
with the education of his son, Alyosha would visit
the house of an old woman to learn to read and write, and
one day, on his way home, he chanced to pass in front of the
Tsar's palace. Well, even well-bred young boys are not
exempt from an occasional fit of curiosity, and Alyosha,
looking about him to ensure that he was not observed, crept
up to the window, stood on the tips of his toes, and looked
in. There in the room sat the Princess, the only daughter of
the Tsar, and you can imagine Alyosha's surprise when,
engaged upon her toilet, the maiden lifted her pretty head
from her shoulders, rubbed it with soap, rinsed it with
water, combed its flowing golden curls, and then put it back
on her shoulders again.

'What a clever woman!' thought the boy, who had never
seen anything like this before. 'She's a real witch!' And
shuddering with fear, he ran swiftly home.

Now although Alyosha did not know it, the Princess, as
soon as she had dried the water from her eyes, had caught
sight of the boy's face at the window, and since it would
clearly not do for people to know that the Tsar's daughter
was a witch, she began then and there to think up all manner

of ways in which to silence him for ever. Before she could put any of her evil schemes to the test, however, she fell grievously ill, and the most she could do in these circumstances was to summon her father the Tsar to her death-bed and get him to promise to appoint the priest's young son to read the Psalter over her corpse for three nights running. The Tsar promised, and the next moment his daughter was dead. They put her body in a coffin and bore it off to the church, where the Tsar asked the priest if he had a son.

'I have one son, Your Majesty,' replied the priest.

'Then command him to read the Psalter over my daughter's body for three nights running,' said the Tsar. 'It was my daughter's dying wish.'

Returning home, the priest told his son to prepare himself for the task. Alyosha said nothing, but later, whilst he was sitting at his books at the house of the old woman, she noticed that his face was pale and drawn and that he paid scant attention to his work.

'What is the matter with you today, Alyosha?' she asked. 'Why are you so unhappy?'

'How should I not be unhappy?' sighed Alyosha. 'I am doomed.'

This was clearly not the language one expected to hear from the mouth of a young boy, so the old woman took his hand and asked him to explain what he meant. Hereupon Alyosha burst into tears and told her all that he had seen at the palace and how the dead Princess was a witch.

'And now I have to read the Psalter over her dead body for three nights running!' he sobbed. 'She will certainly kill me!'

'Do not fear, Alyosha,' said the old woman, who had known all about the Princess. 'I'll tell you what to do. Take this knife, and as soon as you enter the church tonight, describe a circle round yourself and do not move outside it till dawn. Read the psalms aloud, and whatever happens do not look behind you, or you will surely perish!'

That same evening the lad went to the church, described a circle about him in accordance with the old woman's instructions, and began to read the psalms in a loud voice.

When the hour of midnight struck, the lid of the coffin stirred and creaked, and suddenly the witch sat bolt upright in her box, her wild eyes staring. Then, rising slowly to her feet, she groped her way towards the priest's son, the blind white orbs of her eyes rolling in her head, and her sharp teeth gnashing fearsomely. As her bare feet shuffled over the cold stones of the church Alyosha felt his hair bristle on the back of his neck. His heart pounded in his breast and the pages of the book swam before his eyes, but somehow he read on and never looked behind him. When the witch reached the edge of the circle, she came to a sudden stop, and try as she might, she could not cross this line. All night long she clutched and clawed the air at the edge of the circle and the boy seemed to feel her fingers brushing against the back of his head. At dawn, however, the cock crew, and the terrible witch ran back to her coffin and tumbled in. Alyosha's watch came to an end and he returned thankfully home.

The next evening the same thing occurred. At the stroke of midnight the lid of the coffin lifted and fell open, and the witch, stronger now than the night before, leapt out with claws outstretched in an attempt to grasp the boy by the throat. At the edge of the circle, however, she stopped short as though impeded by an invisible wall, and try as she might, she could not penetrate the circle. The priest's son read his psalms aloud, but this time his voice was not alone in the church, for the corpse's mouth began to move, and strange and terrifying words fell from her bloodless lips. A whirlwind blew through the church, and the stone walls echoed to the sound of many creeping, crawling, rustling things. Innumerable pairs of leathery wings seemed to flutter thick around the magic circle, and many claws scratched on the stone slabs of the floor. The boy felt that a host of beetles were crawling over his flesh, but he gritted his teeth and prayed to God, and read his psalms in a loud voice and never once looked behind him. At dawn the cock crew, the witch flung herself into her coffin, and peace reigned again within the church.

When his watch came to an end, the boy went trembling
to the old woman and told her of the horrible things he had
felt around him in the church.

'Alas, boy,' said the old woman. 'There is worse to come!
The third night will be the most trying of all, and you must
be brave. Take this hammer and these four nails, and when
you enter the church, knock a nail into each corner of the
coffin-lid and keep the hammer in front of you whilst you
read the Psalter.'

That evening Alyosha went again to the church and did
exactly as the old woman had instructed him. The hour of
midnight came, the coffin-lid fell to the ground, the body
of the witch shot out and began to run from one side of the
church to the other, conjuring up all the denizens of Hell.
A great wind blew through the church, the holy icons of the
Saints and the Mother of God crashed to the floor, the
windows smashed, and all around the circle in which the boy
stood clamoured a host of evil things with clotted hair,
dragons' tails, bats' wings, crabs' claws, and scorpions' feet.
A great wave of heat passed through the church and lurid
flames seemed to envelop the walls. The priest's son was
terrified, but still he chanted one psalm after another in a
loud voice, and never once did he look behind him. Then,
after what seemed an eternity of torment, the friendly cock
crew, the witch rushed to her coffin, and all the flames of
Hell faded before the first rays of the golden sun. And
there in the open coffin, exposed and face downwards in
ignominy, lay the evil body of the witch.

When Alyosha's watch came to an end, the church door
opened and the Tsar entered. When he saw the state of
his daughter's coffin, he asked the priest's son to explain
what had happened, and hearing that the Princess his
daughter was a witch, the Tsar ordered an aspen stake to be
driven through her black heart, and her body was buried
thus beneath the earth, never to rise again.

And the Tsar rewarded Alyosha, the priest's brave son,
with untold treasure and bountiful gifts.

The Fool and the Birch-Tree

I N a certain kingdom, in a certain land, there once lived
an old man who had three sons, two of whom were
mentally alert, while the third was a fool. When the old
man died, his sons threw lots to divide the patrimony between them and, of course, the clever ones contrived to
secure all the best things for themselves while their foolish
brother got nothing but a bony old ox.

When the day of the local fair arrived and the clever
brothers prepared their wares for market, the fool, not wish-
ing to miss the excitement, decided to do the same, and
tying a rope to the ox's horns, he led it off to town. On the
way, however, he chanced to pass through a wood, and in
that wood stood a withered old birch-tree. The wind blew
and the birch-tree creaked.

'Aha!' said the fool to the tree. 'So you are making a bid
for my fine ox. Well, I am not unwilling to sell it, but its
price is twenty roubles.'

The wind blew and the birch-tree creaked.

'What?' said the fool. 'You want to take it on credit? All
right, then, I'll wait till tomorrow for the money!'

And so, tying the ox to the birch-tree, which gave—or so
he thought—a final grunt of satisfaction, the fool hurried
triumphantly home.

'Well, stupid?' said his brothers. 'Did you sell your old ox?'

'I did.'

'How much did you get for it?'

'Twenty roubles.'

'Good Lord!' exclaimed his brothers. 'Show us the money!'

'I have not got the money yet,' replied the fool. 'But it's all settled—I get the money tomorrow.'

'Blockhead!' cried his brothers, and shrugging their shoulders, they left him.

The next day the fool rose early and went to the birch-tree to collect his money. When he entered the forest, however, the birch-tree stood trembling in the wind, but the ox had gone. Wolves had come during the night and eaten it up.

'Well, neighbour,' said the fool, 'give me the money you promised!'

The wind blew and the birch-tree creaked.

'What do you mean, tomorrow?' protested the fool. 'You said I should have my money today! Very well, I shall wait another day, but no longer, mind! I need the money.'

And leaving the birch-tree, he returned home.

'Well, have you got your money?' asked his brothers.

'Not yet. I must wait a little longer.'

'To whom did you sell the ox?'

'To the withered birch-tree in the wood.'

'Imbecile!' cried the brothers, and they left him in despair.

The next day the fool rose, took his axe, and went to the forest.

'Give me my money!' he said to the birch-tree. 'My brothers know your bad reputation.'

The wind blew and the birch-tree creaked.

'No, neighbour!' said the fool firmly. 'I shall wait no longer. Give me my money or I shall hack you to pieces!'

The wind blew, the birch-tree creaked, and the fool lifted his chopper and began to hack away at the birch-tree until the chips flew in all directions. Now the old tree had a hollow trunk, and in this a band of robbers had hidden a large pot of gold, so that when the tree finally tumbled down, the fool's eyes fell upon a heap of shining gold. The fool was not such a fool as not to recognize *this* for what it was, and gathering as much of it into his skirts as he could, he ran quickly home.

'Where did you get all that gold?' asked his brothers, much bewildered.

'Oh, a little persuasion on my part and the birch-tree soon paid up!' said the fool. 'But I had to leave most of it behind. Let us go and fetch the rest together!'

The three of them ran quickly off to the forest, gathered up the gold, and made their way homewards with it.

'Try to use a bit of sense this time, fool!' pleaded the clever brothers. 'Don't tell anyone we have got the gold.'

'Of course not!' said the fool.

Before they had gone very far, however, they met a deacon, and deacons, as everybody knows, are very inquisitive and greedy fellows.

'What have you got there?' he asked, when he saw the bags they carried.

'Mushrooms!' said the clever brothers.

'What a wicked lie!' exclaimed the fool. 'It's gold. Look!'

And he opened the bag he was carrying and revealed the gold to the deacon's inquisitive eyes. When the latter saw all this gold, he gasped with amazement, fell upon the fool's sack, and began to stuff his pockets full in feverish haste. The fool did not like this behaviour one little bit, and angrily lifting his chopper, he hit the deacon a blow on the head that sent the fellow post-haste to the other world.

'Fool!' cried his brothers in great alarm. 'What have you done, you lunatic? You've ruined yourself, and you'll ruin us too! What are we going to do with the dead body?'

They thought and thought, and finally for want of anything better they dragged the body into an empty cellar and left it there. In the night, however, the clever brothers had grave misgivings about the wisdom of this procedure, and the one brother said to the other:

'This affair might very well turn out badly for all of us, for as soon as people begin to talk about the disappearance of the deacon, our foolish brother will blab everything out. We'd better bury the deacon somewhere else and bury a goat in his place!'

And unbeknown to the fool, this they did.

A few days passed, and the whole village began to search after the missing deacon.

'What are you looking there for, you silly people?' exclaimed the fool. 'Why didn't you ask *me* where to find the deacon? I killed him with my chopper, and my brothers buried him in the cellar.'

The enraged villagers seized the fool and ordered him to show them where the deacon was buried. The fool led them to the cellar, climbed down, and fumbled around in the darkness until his hand fell upon the head of the goat.

'Had the deacon a lot of hair?' he called up.

'Yes.'

'Did he wear a beard?'

'Yes.'

'Did he have horns on his forehead?'

'Horns on his forehead?' cried the villagers. 'What are you talking about?'

'Look for yourselves,' said the fool, and threw them up the head of the goat.

And when the villagers saw that the body in the cellar was that of a goat, they spat in the fool's eye and went off home.

Fair Vasilissa and Baba Yaga

Long, long ago in Old Russia, a rich merchant married a beautiful young wife, and when they had been married for four years, she bore him a fine golden-haired, blue-eyed daughter whose beauty quickly earned her the flattering name of Fair Vasilissa. When the child was only eight years old, however, the mother fell gravely ill, but before she died, she summoned Vasilissa to her bedside, took a tiny doll from beneath the bedclothes, and gave it to her daughter.

'Fair Vasilissushka,' she murmured, 'you must know that I am dying and have called you here to bestow my last blessing upon you. Take also this tiny doll as one last present. Show her to no one, and if some misfortune befall you, give the doll food and she will tell you what to do.'

After the death of his fair wife, the merchant mourned deeply, but the passage of time and the cares of business eventually healed the wound, and he began to think of taking another wife. It was not long, therefore, before a certain small widow caught his fancy. The woman was no longer young and had two daughters of her own who were a little older than Vasilissa, but the merchant, who was a good man, thinking that she would a good mother for his own child, married her. He proved to be sorely mistaken in

his calculations, however, for his new wife was not at all kind
to Vasilissa. In the whole village there was no girl more
beautiful than Fair Vasilissa, and this state of affairs made the
stepmother and the stepsisters wax green with envy, and
they took to giving her the most difficult tasks to perform
in the hope that her body would grow thin and bony and her
fair white face blackened by the sun and torn by brambles.
But however assiduously they worked for Vasilissa's ruin, it
was they themselves that grew black, thin, and ugly, while
their fair stepsister grew up into a tall, firm-limbed, full-
bosomed maiden. It must be noted, however, that had it not
been for the magic doll who performed all the heavy and un-
pleasant tasks for her mistress, weeding the garden, filling
the buckets, watering the cabbages, and heating the stove,
while Vasilissa dallied in the shade and made garlands of wild
flowers, their wicked schemes might well have succeeded.

As the years passed and Vasilissa grew up to maidenhood,
there was not one eligible youth in the village who did not
desire her for his bride. No one gave a second glance at
her ugly stepsisters, of course, and when the stepmother
received the suitors' matchmakers, she would grow exceedingly angry.

'The younger sister shall not marry before her elders!' she
cried, and when the suitors had left the house, she would
beat Vasilissa out of pure spite.

Well, the day came when the merchant was obliged to
leave home on a prolonged business journey, and in his absence the evil stepmother moved the family to a small house
at the edge of a forest. In the depths of that dark forest was
a clearing, and in the clearing stood a hut on fowls' legs, and
in that hut dwelt Baba Yaga the witch, who with her sharp
and greedy teeth devoured every unfortunate human who
chanced to stray within her reach. Only too aware of this,
the evil stepmother seized every possible opportunity of
sending Fair Vasilissa on errands into the dreaded dark
forest near Baba Yaga's hut, but invariably the tiny doll
would show her mistress the way to avoid the dwelling of the
foul old witch.

The long dark evenings of autumn came, and on one of these the evil stepmother allotted various tasks to her daughters and stepdaughter: one daughter was to make the lace, the other was to knit the stockings, and Fair Vasilissa was to weave the cloth. When darkness had fallen in the forest outside, the evil stepmother put out all the candles in the house except the one by which the three girls worked and then went to bed. After a while, the lone candle began to smoke and smell badly, and one of the stepsisters took a pair of snuffers and made as if to crop the ill-burning wick, but, in accordance with the secret instructions of her artful mother, she pretended to stumble and snuffed the candle right out. The whole house was plunged into darkness.

'What shall we do?' wailed one sister. 'There is no fire anywhere in the house, and we have not yet finished our work!'

'Baba Yaga will have a light,' said the other. 'But *I'm* not going!'

'Nor I!' said the first. 'Vasilissa has almost completed her task, so *she* can go and ask Baba Yaga for a light!'

And with that, they pushed Vasilissa roughly out of the room.

Vasilissa went up to her tiny chamber and prepared food for her doll. When the doll had eaten, her eyes began to glow like candles, and when her young mistress told her of her present predicament, she comforted her.

'Don't worry, Vasilissushka!' she said. 'Let us go to Baba Yaga, and I shall see that no harm befall you.'

So placing the doll carefully in her pocket and putting on her shawl, Fair Vasilissa crossed herself and walked out into the dark forest.

Suddenly the earth began to tremble, the trees shook, and she heard the sound of beating hooves behind her. She barely had time to dart into a thicket at the edge of the road before a strange horseman galloped swiftly past: the rider was white, the horse was white, and both were accoutred all in white. And as he passed by, dawn broke.

Vasilissa went her way, somewhat heartened by the light

of day that filtered through the thick forest trees, when suddenly she heard again the sound of galloping hooves. Again she hid in the bushes and saw another strange horseman ride swiftly by: the rider was red, his horse was red, and both were accoutred all in red. And as he passed by, the sun rose in the heavens.

Warmed by the heat of the beneficent sun, Vasilissa walked on and on through the forest all day, and eventually she arrived at the little clearing in the forest where the hut of Baba Yaga stood on its fowls' legs. To her horror, Vasilissa saw that the fence round the hut was of dead men's bones and that on each weird stake was stuck a skull with staring eyes; the posts of the gates were dead men's legs, and the lock was a mouth with long, sharp teeth!

Vasilissa was at first petrified with fear, but suddenly she heard the sound of beating hooves behind her, and hiding once more in the bushes, she saw another horseman gallop swiftly past: the rider was black, his horse was black, and both were accoutred all in black. And riding up to Baba Yaga's gate, he disappeared as though the very earth had swallowed him up, and dark night fell. And in that very moment the eyes of the dead skulls on the fence of dead men's bones lit up like lamps, and the clearing shone as bright as day.

Vasilissa was already trembling with fear, but worse was to come, for suddenly a terrifying sound issued forth from the forest : night-birds screamed, the trees crackled, the dry leaves rustled, and out of the dark forest, riding in a mortar, propelled by a pestle, and sweeping away her traces with a broom, came Baba Yaga the witch herself!

And she was as sinewy, gnarled, and bloodless as the roots of a thunderstruck tree!

At the gate of her hut she paused and sniffed the air.

'Faugh! Faugh!' she cackled gleefully. 'I smell Russian blood!'

Pale with terror, Vasilissa saw that it was useless to try to conceal herself, and going up to the fearsome witch, she bowed low before her.

'It is I, babushka,' she faltered. 'My stepmother's daughters sent me to ask you for a light.'

'I know them well!' cackled the witch. 'So you are Fair Vasilissa, are you? Well, work for me and I'll give you a light for your trouble. Refuse, and I'll gobble you up!'

And with that, she seized Vasilissa's arm in her bony hand and hustled her roughly towards the hut.

'Divide, firm fence!' she cried. 'Open up, wide gates!'

The strange fence and the gates obeyed Baba Yaga's command, and when the witch had driven in with Vasilissa, they rattled shut again behind them.

On entering the hut, Baba Yaga stretched herself out in comfort on the sofa.

'I am tired and hungry!' she cried. 'Give me to eat whatever there is in the oven.'

Vasilissa lit a wooden spill from the fiery eyes of the skulls on the fence and searched about in the oven. She set in front of Baba Yaga food enough for ten men together with kvass, mead, beer, and wine from the cellar. The witch ate and drank greedily, and all that was left for Vasilissa were a few scraps, a crust of bread, and a piece of sucking-pig.

'When I go forth tomorrow,' said Baba Yaga, lying down to sleep, 'you will wash down the yard, sweep the hut, cook the dinner, and wash the linen; then go to the corn bin, take out a quarter of wheat and clear it of every trace of fennel. And see to it that all is done by the time I return, or I shall gobble you up!'

Whereupon, having said this, Baba Yaga turned over on her side and began to snore loudly.

Vasilissa placed the scraps of food left over by the witch in front of her doll, and bursting into tears, told her what had happened.

'Do not fear, Fair Vasilissa!' said the doll. 'Eat this food yourself, say your prayers, and go to sleep, for morning is wiser than the eve.'

The next day Vasilissa awoke early, but Baba Yaga was already up, and when she looked out of the window, she saw that the light in the eyes of the skulls was growing dimmer and dimmer. Suddenly, the white horseman appeared as if from nowhere and galloped swiftly away. The light in the eyes of the skulls went out completely, and dawn broke. Baba Yaga stepped outside, whistled loudly, and the mortar and pestle and broom appeared before her. Shortly afterwards, when the red horseman appeared and the sun rose in the sky, Baba Yaga sat in her mortar and rode swiftly from the yard, propelling herself along with the pestle and sweeping away her traces with the broom.

When Vasilissa returned from the window, she saw that all the tasks the witch had set were already done: the yard was cleaned, the hut was swept, the linen washed, and the quarter of wheat cleared free of fennel. The doll had done everything.

'You have saved me from great misfortune!' Vasilissa cried gratefully to her doll, as it slipped back into her pocket.

Towards evening, Vasilissa laid the table and waited for Baba Yaga. Twilight grew thicker, the black horseman rode up and all went black. The eyes of the skulls lit up, nightbirds screamed in the forest, the trees shook and the leaves rustled, and Baba Yaga drove through the gates.

'Is everything done?' said the Yaga.

'Look for yourself, babushka,' replied Vasilissa.

When the witch saw that everything had been done according to her instructions and there was nothing with which to find fault, she was secretly annoyed.

'All right!' she snapped, and clapping her hands, she cried:

'My faithful servants, grind my wheat!'

Immediately three pairs of hands appeared from nowhere, grasped the bushels of wheat and dragged them away to be ground.

When Baba Yaga had eaten her large meal, she lay down on her couch.

'Tomorrow you shall do exactly as you did yesterday, but in addition you shall take the poppy seed from the corn bin and cleanse it of every grain of earth.'

And smiling wickedly to herself, Baba Yaga turned on her side towards the wall and began to snore.

Vasilissa gave the doll to eat what little the witch had left over and told it of her new tasks, whereupon the doll again told her to eat and say her prayers and not to worry.

The next day, when Baba Yaga rode back on her mortar, the doll had performed to perfection all the tasks the witch had set Fair Vasilissa, and the witch could find fault with nothing.

'Ho there, faithful servants!' she cried. 'Press the oil from the poppy seed.'

Again the three pairs of hands appeared and bore the baskets of poppy seed away out of sight.

'Why do you never speak to me, girl?' said Baba Yaga, as she sat down to eat. 'Are you dumb?'

'I dared not speak, babushka,' replied Vasilissa. 'But since you invite me to do so, there is something I should like to ask you.'

'Well, ask away!' said the witch. 'But remember, it is not every question that leads to good. The more you know, the older you grow.'

'I only wanted to ask you,' said Vasilissa timidly, 'about the three strange horsemen who gallop to and from your hut, the one white on a white horse, one red on a red horse, and one black on a black horse.'

'The White Horseman is the Bright Day,' replied Baba Yaga, 'the Red Horseman is the Fair Sun, and the Black Horseman is the Dark Night. All three are my faithful servants. Are you satisfied?'

Vasilissa dearly wanted to ask about the three pairs of hands, but she said nothing.

'You have grown dumb again, child!' said Baba Yaga. 'Do you not wish to know anything else?'

'I have asked enough, babushka,' replied Fair Vasilissa. 'You said yourself that the more you know, the older you grow.'

'You are a wise child, Vasilissa,' said Baba Yaga. 'I eat up inquisitive little girls and boys. But now I'll ask you some-

thing. How do you manage to perform the difficult tasks I
set you to do?'

Vasilissa remembered her mother's warning to say
nothing of the doll to anyone.

'My mother's blessing aids me,' she replied.

'What!' cried Baba Yaga. 'Begone then, blessed child!
I want no blessings in my house!'

And clutching Vasilissa roughly by the arm, she dragged
her outside the fence.

'You have performed your service well,' said the witch,
'so you may take what you came for.'

Baba Yaga took one of the shining skulls from the fence,
stuck it on a stick, thrust it into Vasilissa's hand, and told her
to go.

Fair Vasilissa needed no second bidding, and ran away as
fast as she could with the light from the skull to guide her
way home. At dawn the light went out, but it was already

evening again before she arrived at her stepmother's house, and the eyes of the skull began to glimmer again.

'Nearly a week has passed since I left,' thought Vasilissa. 'Surely they no longer need a light at home.'

She was about to throw it away when the skull spoke.

'Do not throw me away, Vasilissa,' it said. 'Take me to your stepmother.'

Now ever since Fair Vasilissa had left, no light or fire could be induced to burn in the stepmother's house. The mother and her daughters could not kindle a light themselves, and as soon as they brought one in from a neighbour's house it went out immediately. It was therefore with some impatience that the wicked stepmother greeted Fair Vasilissa.

'You have kept us waiting long enough!' she cried. 'Bring that light in at once!'

But as soon as Vasilissa stepped over the threshold with her strange torch, great flames shot out of the eyes of the burning skull and enveloped the wicked stepmother and her evil daughters in a cloak of fire, and when there was nothing left of them but three heaps of grey ash on the ground, the eyes of the skull went out for ever.

That same evening the merchant returned from his long journey and Vasilissa told him all that had happened in his absence. Her father was pleased to be rid of the wicked wife who had planned to send his beloved daughter to her death, and soon afterwards the Fair Vasilissa married a handsome youth and all lived happily ever after.

The Frost, the Sun, and the Wind

A PEASANT was walking along a country road when he
met the Frost, the Sun, and the Wind journeying
together in the opposite direction.

'Greetings!' said the peasant, and went his way.

'Which one of us did the peasant greet?' asked the Frost,
after a while.

'Me, of course,' said the Sun, 'so that I should not burn
him up.'

'Nonsense!' said the Frost. 'He greeted me, for none is
so feared by mortal men as I.'

'You are both wrong,' said the Wind. 'It was I he
greeted.'

They quarrelled and argued and almost came to blows,
but they could reach no decision.

'Since there is such dissension between us,' they said, 'let
us catch the peasant up and ask him.'

This they did.

'Whom did you greet, brother?' they asked.

'The Wind,' replied the peasant.

'You shall not forget me, my fine peasant friend!'
said the Sun. 'I shall broil you and cook you with my
rays!'

'Have no fear, friend,' said the Wind. 'I shall blow on you and cool you with my soft breezes.'

'I shall freeze you into a block of ice, peasant!' threatened the Frost.

'Have no fear, brother,' said the Wind. 'There is no Frost, if I do not blow.'

The Angel

GOD sent an angel to take the soul of a woman
who had just given birth to twins, but when the
angel saw the two defenceless newborn babes, he
was filled with pity and returned to Heaven without fulfill-
ing his task.

'Well?' said the Lord God. 'Have you taken her soul?'

'No, Lord.'

'Why not?'

'Lord, the woman had two little children! Who should
look after them if their mother die?'

Thereupon God took his mace and struck a rock with it.
The great rock split in two.

'Go down into this rock!' said the Lord God.

The angel descended into the rock.

'What do you see?' asked God.

'I see two small worms,' replied the angel.

'Then would not he who feeds these worms,' said God,
'have fed the woman's newborn babes?'

And God took away the angel's wings and sent him to
dwell in the world of men for the space of three whole years.

On earth, the angel hired himself as a labourer to a
village priest, and for one year and then another he served
his master faithfully. One day in the third year the priest

sent him on an errand into the town, and on his way the angel, chancing to pass by a church, looked up and suddenly began to hurl stones at the Cross on the dome of the church. A crowd gathered, and angrily falling upon the angel, they beat him severely. When he came to his senses, the angel got up and walked on until he came to a tavern, and entering the door, he knelt down and began to pray.

'There's a fool for you!' said the passers-by. 'He throws stones at the church and prays in the tavern. He ought to be whipped!'

But the angel crossed himself silently, rose and went on his way until he came to a beggar, whereupon he began to upbraid the man as a wheedling thief. When the villagers saw this, they seized the angel, beat him again, and dragged him back to the priest his master.

'Priest,' they said, 'your hired man goes through the streets throwing stones at the church, mocking God in the tavern, and ill-treating the poor and needy. Is such a blasphemous and uncharitable fellow fit to be your labourer?'

'What have you to say to those charges, brother?' said the priest.

'As I was passing the church,' said the angel, 'I spied the Devil, emboldened by Man's many sins, squatting on the roof of the church with his arm round the Cross, so I drove him away with stones. As I passed the tavern, I saw many people inside drinking, quarrelling, and blaspheming, heedless of the hour of their death, and I knelt to pray to God to save all true believers from eternal perdition.'

'But why did you ill-treat the poor man?' asked the priest.

'It was no poor man,' replied the angel, 'but Dives in the rags of Lazarus. The man has much money, but he goes about the land asking for charity and taking the bread from the mouths of the truly poor, and that is why I called him a wheedling thief. But now, priest, I have served you well for three years. I have no need of money, but as a reward for my services I should like you to walk with me awhile.'

The priest went with the angel, and they wandered out into the open countryside. And as they walked together, the priest suddenly saw that the face of his labourer shone with a heavenly light and that from his shoulders a pair of white and shining wings had issued forth, but before he could speak a word, the skies opened, and the angel rose from the earth and ascended to Heaven robed in glory. And the priest knew his hired labourer to be a messenger of the Lord.

The Magic Berries

IN a certain kingdom, in a certain land, there lived a
King and a Queen and their daughter the Princess.
The Princess was a very beautiful maiden, and although
she had rather more than her fair share of curiosity, greed
and pride, her royal parents paid no attention to these faults,
from which, to tell the truth, they were not entirely exempt
themselves, and they cherished their daughter like the apple
of their eye.

Well, on the day our story begins, there sailed into the
harbour of the royal city a splendid ship from foreign parts,
and all the townsfolk great and small crowded on the quayside to watch. As soon as the ship had docked, the owner, a
rich merchant whose portly form was attired in the most
wonderful silks and brocades, stepped ashore and ordered
his men to display all manner of rare and wondrous things
to the admiring eyes of the gaping onlookers. The riches of
China and Levant were not unknown to the subjects of this
kingdom, but, to tell the truth, this merchant had things
that no one, not even the oldest grandfather, had ever seen
before, and the news of his arrival quickly spread to the
palace.

The Princess, whose curiosity, as you can imagine, was

174

ever ready to be awakened, clapped her hands for joy, and most eager to see the reputed wonders for herself, she begged her father and mother to give her permission to descend to the sea-front just this once. The King, who was loth to let his daughter out of his sight for an instant, tried to persuade her that popular rumour distorts all things, but the Princess pouted and pleaded so prettily that he finally gave a grudging consent.

'But mind you look after the Princess!' he warned the governesses and handmaids of her suite, wagging his finger severely. 'If anything happens to her, I'll cut all your heads off, every one!'

When the Princess arrived with her suite on the quay, the rich merchant welcomed her warmly, and showed her the many rich and wonderful wares he had laid out on the ground with the apparent object of dazzling the townsfolk and inducing the merchants of the kingdom to part with their money. The Princess was delighted to see such riches, and she asked how much this was, and how much that was, until, had she actually purchased all that took her fancy, she would have quite impoverished the Royal Exchequer. The merchant, however, seeing the gratifying effect his goods had produced on the Princess, took her gently aside and whispered in her ear.

'Fair Princess,' he said, 'all this is as nothing and quite unworthy of a royal personage such as you. If you really want to see things wonderful, miraculous, and out-of-this-world, come aboard my ship. I will show you a talking cat, a self-playing lute, a self-laying table-cloth, and a host of other wonders of the world. These things I have shown no one, for I have reserved them all for you.'

The Princess, however, though flattered, was a little uneasy and made to go away, but the merchant reassured her craftily.

'Fair Princess,' he said, 'do not go away. Come aboard my ship for two minutes only, and if any of my possessions take your fancy, you shall take them back to your palace as a gift.'

Who can resist the persuasion of a wily merchant? The Princess's curiosity overcame her timidity, and commanding her suite to await her return on the quay, she accompanied the merchant aboard his ship. The merchant led her to a most luxurious cabin and bade her sit and wait there while he went to fetch the magic wonders of which he had spoken. No sooner had he left the cabin, however, than he slammed the door and locked it securely behind him.

'Raise the anchors!' he yelled, and in a trice his crew scaled the rigging, the sails were hoisted, and within seconds the ship was sailing away into the open sea.

'Help! Help!' wailed the governesses and handmaids on the quayside. 'The wicked merchant has kidnapped our young mistress!'

But however much they wept and lamented, the ship grew smaller and smaller in the distance until it finally disappeared over the horizon. By the time the King and the Queen drove up, it was much too late. The ship and the Princess were far away to sea.

The Queen—such is the way of women—fainted on the spot, but the King—such is the way of men—grew purple with anger, ordered all the governesses and handmaidens to be cast into the deep dungeons, and bade the herald announce to all people far and near that whosoever should rescue the fair Princess and return her safe and sound to her royal parents should receive her hand in marriage and should rule the whole land after his death.

Many were the volunteers, and many were the failures, for though innumerable young and brave men sought the Princess to the far corners of the earth for the space of three whole years, she was nowhere to be found, and the people began to murmur that she had been spirited away. In the royal city, however, there served a Soldier called Ivan, the son of a peasant, and one evening it fell to his lot to keep watch over the King's gardens. Ivan stationed himself beneath a tall tree, and being a conscientious fellow out of the common run of sentries, he kept himself awake and watched carefully that no intruder should penetrate the King's property. At midnight, two black crows flew up and settled on one of the branches of the tree where Ivan stood guard.

'The King of this country has lost his only daughter,' squawked the one. 'They have sought her everywhere for three years, and no one has ever found her.'

'What fools these humans are, to be sure!' cried the other. 'They had only to sail to the warm Southern seas to the land of Nemal Chelovek,* for he it is who kidnapped the fair Princess in the guise of a merchant and now holds her captive in his palace with the intention of giving her in marriage to his nephew, the Dragon Gorynich. Finding the Princess is easy, but none could ever escape alive, for Nemal Chelovek is invincible.'

'That is not so,' objected the first crow. 'Even Nemal Chelovek is vulnerable. There is a tiny island in the middle of the ocean where two leshis † dwell. They pass their days in perpetual disharmony, for for thirty years they have been squabbling over the possession of the Samosek Sword.‡ If only there were a man in the world brave enough and clever enough to steal the sword from these two leshis, he could easily overcome Nemal Chelovek!'

Hereupon the two crows flew off. Ivan did not delay. As soon as his watch was over, he went to the palace and asked to see the King.

'What do you want, Soldier?' asked the latter, when the Soldier was finally admitted into his presence.

'I want permission to go in search of the Princess your daughter, Your Majesty,' said Ivan.

'What!' cried the King, renowned more for irascibility than patience. 'You? Better men than you by far— princes, boyars, merchants, and generals—have already sought the Princess to the ends of the earth, and not one has succeeded in the quest! How can you, a private soldier, the son of a peasant, hope to succeed where so many have failed?

* The 'Big Man'.   † Woodland Spirits.
‡ Self-cutting Sword.

You have been nowhere and seen nothing. I warn you, Soldier, if this is a trick to get leave from your regiment, you shall rue the day you thought of it, for if you fail to find the Princess now, your head shall surely fall!'

'Your Majesty,' replied Ivan calmly, 'I *know* where to find the Princess.'

The Soldier's quiet assurance impressed the King, and he rubbed his chin thoughtfully.

'Very well, Soldier,' he said at last. 'If you find the Princess, she shall be your wife and the Kingdom shall be yours to rule after my death. That is a bargain, and I shall stick to it. If, however, you fail to find her, I shall put you to death for your insolence!'

'A man cannot die two deaths, Your Majesty,' said the Soldier, shrugging his shoulders, 'and no man shall avoid dying once. Order a ship to be got ready and command its captain to obey me implicitly.'

Within a day or so, therefore, the Soldier was sailing across the deep blue sea towards the island where the leshis dwelt.

'Drop anchor here, Captain,' said Ivan, as the coast loomed before them. 'I shall go ashore. But stand ready, and as soon as I return, raise the anchor, hoist the sails, and retreat from this island at full speed!'

Ivan the Soldier leapt on to the steep and rocky shore, and directing his steps inland, he walked on until he came to a thick forest. All of a sudden a great noise of breaking twigs and shouting came to his ears, and out of the wood rushed the two leshis, fighting furiously for the possession of something they strove to tear from each other's grasp.

'Let go! I shall never give it up!' shouted one.

'Give it here, it's mine!' cried the other, and both fell in a struggling heap on the ground. When they became aware of the Soldier's presence, they stopped fighting and appealed to the Soldier.

'Be our judge, good man!' they said with one breath. 'Thirty years ago we inherited this sword between us, and ever since that day we have been fighting and squabbling for its possession, for we are two, but the sword, alas! is only one!'

'If it's agreement you're after,' said Ivan, 'that should present no difficulty. Give me the sword to hold. I shall shoot an arrow into the wood, and he who brings back the arrow shall have the sword.'

Dwelling in a wood does not sharpen the wits, and the two leshis readily agreed to the Soldier's proposal. Ivan took the magic Samosek Sword and shot an arrow into the wood. The leshis dashed off as fast as their legs would carry them, and within seconds they had pounced on the arrow and were struggling as bitterly for its possession as they had done for the Samosek Sword. Ivan wasted no time. He ran back to the ship, and in little more time than it takes to tell, Ivan and his ship were sailing the high seas with the island of the quarrelsome leshis far behind them.

For a day and a night they sailed on and on, and when the sun rose the next morning and shone on the blue waters of the Southern seas, they caught sight of the land of Nemal Chelovek. Ivan gave his orders to the captain of the ship and went ashore in search of the lost Princess. Not far from the shore stood a large white mansion, and finding it un-guarded—for Nemal Chelovek never conceived the possibility of any ordinary mortal being rash enough to trespass on his domains—he opened the door and went in. And there in the corner, weeping her eyes out, sat the Princess.

The brusque entry of the unknown traveller startled her, and she looked up fearfully.

'Who are you?' she gasped. 'How did you get here?'

'I am a Soldier, my name is Ivan,' he replied. 'I have come to take you home to your parents.'

The Princess was reassured by his words that he meant her no harm, but no flicker of hope showed on her face.

'Alas, Soldier,' she sighed. 'No one can escape from this island, and Nemal Chelovek will certainly kill you when he returns!'

'We shall soon see who kills whom!' said the Soldier grimly, grasping the hilt of the Samosek Sword, and his tone was so encouraging that the Princess dried her eyes and smiled a wan smile.

'Soldier,' she said, 'if you deliver me from this monster Nemal Chelovek and his terrible nephew the Dragon Gorynich, I will gladly marry you.'

'So be it,' said Ivan. 'But will you keep your promise when the time comes?'

The beautiful Princess removed the golden ring from her finger and handed it solemnly to the Soldier.

'We are hereby betrothed,' she said. 'I shall keep my word!'

Hardly had these words left her lips than a great clamour echoed through the halls of the palace as though Behemoth and Leviathan themselves had entered the door together.

'Quick! Hide yourself, Soldier!' exclaimed the Princess. 'Nemal Chelovek is coming!'

Ivan ran quickly behind the stove. The door burst open, and into the room stepped the gigantic form of Nemal Chelovek. His head brushed the ceiling, his huge frame shut out the light of day and cast a dark shadow across the room. On the threshold he paused and sniffed the air.

'Faugh! Faugh! Faugh!' he roared, and the rafters shook at the sound. 'It is many years since I was in Russia and smelt Russian blood! Come out from behind the stove, you cowardly, quivering, snivelling bogatyr, that I may clap you between the palms of my hands and squash your puny body to dust and water!'

'You boast too soon, Nemal Chelovek!' cried the Soldier, angered by the taunts. 'The funeral wake shall be in your honour, not mine!'

And saying thus, he brandished the Samosek Sword aloft, the keen blade whirled through the air, and the mighty head of Nemal Chelovek rolled across the marble floor. The giant's guards rushed up to avenge their master, but the magic sword span through the room and hacked them all to mincemeat. When all the fighting was done, Ivan took the Princess by the hand and led her down to the waiting ship.

The winds blew, the sails swelled round and full, and the happy band sailed safely back to their native land.

When the Princess and Ivan arrived in the royal city, the King and Queen laughed and wept for joy and embraced their newly-found daughter with heartfelt sobs, while the whole kingdom glorified the manly valour of Ivan the Soldier. A great banquet was held, and all ate, drank, and made merry to the sound of lute and pipe and drum.

'Ivan, son of a peasant!' cried the King. 'You were once a private soldier, but now, in recognition of your great services, I promote you forthwith to the rank of general!'

'Thank you, Your Majesty,' replied Ivan; 'but are we not forgetting something of greater moment? An agreement is a sacred thing, O King, and I think it time to prepare for the wedding.'

The King blushed with embarrassment.

'I remember our agreement, Soldier,' he said. 'But since then some unfavourable circumstances have developed and it is no longer so simple a matter as before. A foreign prince has asked me for the hand of my daughter. He is a good match, and I should not like to compel the Princess to marry you against her will. Shall we not let her decide for herself?'

'She has already decided, Your Majesty,' said Ivan, showing the King the golden ring the Princess had given him in the palace of Nemal Chelovek. 'We are already betrothed.'

Now the King was exceedingly loth to ally himself or a member of the Royal Family with a commoner, especially the son of a peasant, and to reject the suit of so suitable a foreign prince went hard against the grain. But, as Ivan said, an agreement is a sacred thing and the King's word must be his bond.

'Very well, Soldier,' sighed the King. 'If the Princess has already given you her hand, the wedding shall be celebrated forthwith!'

And so Ivan the Soldier married the beautiful Princess. The story, however, does not end there, for the marriage feast had not yet reached its end when a messenger burst into the hall with the news that the foreign prince, furious at the rejection of his suit, had invaded the kingdom with a powerful army and intended to avenge this slight on his honour.

'If you do not get rid of this Ivan fellow,' he threatened the King, 'and give me your daughter's hand, I shall subject your lands and citizens to slavery, pillage, and arson!'

The King and his boyars were terrified, and the Princess herself fell prey to very mixed feelings.

'Alas!' she thought to herself. 'If only I had not given my hand to this son of a peasant so hastily, I should have married this manly and well-born foreign prince and my parents would now be free from care.'

When Ivan beheld the depressed mien of his young wife and the undisguised anxiety of the whole assembly, he guessed the thoughts that must be passing through everybody's mind, and rising from the banqueting table, he mounted his horse, rode out alone against the enemy hordes, and fell upon them with his Samosek Sword. The magic weapon whirled aloft, horse and infantry were mown down like grass, and within minutes the whole army was utterly destroyed. Only the enemy prince and his generals escaped, and Ivan, returning in triumph from the field of battle, was given a great reception by the grateful citizens. Joy filled all men's hearts and there was no sad face in all the land. But though the Princess laughed with the rest, her happiness was only simulated, for her heart grieved within her.

'Now I shall have to spend the rest of my life with this son of a peasant!' she said. 'Why was he not killed?'

Some time passed, and the news came to the King that the foreign prince had formed a new army and threatened to enslave the land and to carry off the Princess by force. Ivan did not hesitate. He leapt on his trusty steed, brandished his Samosek Sword aloft, and again the foreign host was mown down like grass. This time the prince and his generals barely managed to escape with their lives. Ivan's victory was again fêted by his thankful compatriots and he did not suspect the resentment that rankled in the Princess's heart.

The foreign prince, however, who knew of the decidedly
haughty nature of the Princess, decided to enlist her aid in
his hitherto unsuccessful struggle to make her his wife, and

as soon as he arrived back in his own country, he sent her a
secret message.

'It is not possible,' he wrote, 'for one man to slay a whole
army unless he be possessed of some magic power. Question
your husband and discover his secret, for if I win the next
battle you will one day be Queen. Should I fail again, however, you will remain the wife of a peasant for ever!'

The Princess, whose vanity and ingratitude had blinded
her to the debt she owed Ivan, who, if he was a peasant, was
as a husband preferable to the Dragon Gorynich from whom
he had saved her, resolved to collaborate with the foreign
prince. And so, on the first occasion that presented itself,
she put her arms around the Soldier's neck and told him all
manner of flattering lies.

'Tell me, Ivanushka,' she said in a velvety voice calculated
to undermine the resistance of the most single-minded of
men, 'why did you never tell me how you managed to slay the
giant Nemal Chelovek and defeat those two great armies all
alone?'

'I possess the leshis' Samosek Sword,' replied her husband, quite unconscious of the danger that threatened him. 'Whoever possesses this sword shall defeat the mightiest bogatyrs and return unscathed from the fray, for it cuts down men of its own accord.'

The Princess was well content with this information, and the very next day she went to the Royal Armourer and bade him fabricate a sword identical to that of her husband. When it was finished, she removed the Samosek Sword from its scabbard and placed the counterfeit in its place. Then she sat down and wrote a letter to her foreign suitor.

'Assemble your armies and attack!' she wrote. 'There is nothing to fear!'

Within days the news of an immense advancing army reached the King and threw him into a fresh panic. Again Ivan the Soldier grasped his sword—or what he imagined to be his sword—leapt on his horse and rode out against the foe. This time, however, although he fought valiantly and wounded many, he was quickly overwhelmed by sheer weight of numbers, stricken from his horse, and left for dead on the field of battle with grievous and bloody wounds. The foreign prince was triumphant, the Princess was radiant with joy, and the King quickly adapted himself to the new situation and was relieved at the thought of final (albeit ignominious) peace. The foreign prince wedded the Princess and the marriage was celebrated in great pomp.

When night fell over the battle field, the cold wind brought Ivan to his senses, and he awoke to find himself surrounded by the many corpses he had slain. He crawled painfully to the edge of the plain and hid himself in the thick forest, where he washed his wounds in the cool brook and bandaged them with leaves. As soon as the worthless sword was slashed from his hand, Ivan had realized the treachery of his wife, and now, tired and hungry, he wandered through the forest in search of food and drink. He had not gone far, however, when his eyes fell upon a bush covered with ripe yellow berries, and although he did not recognize them, he was too hungry to care. He picked one of the berries and ate it, and since the taste was good, he picked another and ate that. Suddenly his head began to ache and throb madly at the temples, and when he placed his hand on his forehead to soothe the pain, his fingers caught on a rough and hard protuberance. A pair of horns had sprouted from his forehead!

Ivan's soul touched the depths of despair.

'Alas!' he sighed. 'It seems my misfortunes will never end, for now I can no longer show myself among ordinary folk and shall have to live in the forest for ever!'

He wandered on a little farther until he came to another small bush. It was much like the first, but this one was covered in ripe *red* berries. He did not recognize this species either, but he was still hungry and thirsty, and what with the horns sprouting from his forehead already, he did not much care what happened to him. He stretched out his hand, plucked one of the red berries, and ate it. Immediately one of the horns fell from his forehead to the ground. He ate another, and the second horn fell also to the ground, and he felt that his strength had increased threefold within him.

Now that he was well again, Ivan's first thought was to retrieve his lost Samosek Sword. He exchanged his flowered court garments for a ragged old smock, and filling one small basket with the yellow berries and another with red berries, he made his way to the royal palace.

'Ripe berries!' he called. 'Sweet ripe berries!'

Now the Princess, along with her other faults, was an exceedingly greedy girl, and when she heard the beggar calling out his wares in the courtyard, her mouth watered so much that she dispatched her maid to bring her some berries.

'Are the berries sweet?' the maid asked Ivan.

'You will find no sweeter berries in the whole wide world,' replied Ivan. 'Try one!'

And he gave her a large red berry.

The maid, who had been rushing around at her mistress's beck and call all morning and was beginning to feel tired, suddenly felt all her fatigue vanish from within her as soon as the red berry had passed her lips, and she eagerly bought a basket from the supposed beggar. This time, however, Ivan sold her only the yellow berries, and it was accordingly these that she brought back to the Princess.

'You have never tasted berries like these,' said the maid, not knowing how true this was, when she came back to her mistress.

The Princess took first one yellow berry and then another, when suddenly her head began to ache.

'Oh, how my head aches!' cried the Princess.

The maid looked at her mistress, and seeing the two great horns sprouting from her forehead, was so aghast she could not utter a word. The Princess ran to her mirror.

'Help! Stop the beggar!' she screamed. 'He has bewitched me!'

But Ivan had long since disappeared, and was nowhere to be found.

Many doctors came to cure the sick Princess, but none of their remedies helped her in the slightest. The horns still grew on her forehead, and the Princess began to waste away with worry and shame. In the meantime Ivan the Soldier grew a long beard, and disguising himself as an old, old man, went before the King.

'Your Majesty,' he said, 'I have a medicine to cure all illnesses, and I have come to make your daughter well again.'

The King was overjoyed.

'Cure my daughter of this dreadful affliction,' he said, 'and you may ask of me what you will!'

Ivan had heard the like of that before.

'Thank you, Your Majesty,' he said, 'but I desire no reward. Just lead me to the Princess and promise me that no one shall intrude upon us until I myself call. If the Princess calls out, it will be because the cure is a little painful, but let none enter. If you cannot promise me this on your royal honour, the Princess will never lose her horns.'

The King was only too glad to agree to these conditions, and leading Ivan to the Princess's chamber, he left him there with her and gave the servants strict instructions not to enter for any reason whatsover unless the doctor himself summoned them.

Alone with his false wife the Princess, Ivan locked and relocked the door, and drawing a birchwood cane from his bag, he began to beat the Princess mercilessly. The Princess screamed and shouted for help, but all to no avail. The cane was of birchwood, not of alder, and it did not bend and it did not break, and it beat the Princess black and blue.

'Let that be a lesson to you!' panted Ivan. 'In future you shall deceive nobody by your perfidy!'

When the Princess saw that the doctor was none other than her husband Ivan the Soldier, she screamed even louder than before and appealed frantically for help. But the servants obeyed their master the King to the letter, and no help came, and Ivan beat his wife even harder.

'If you do not give me back my Samosek Sword,' he said, 'I shall kill you!'

'Don't kill me, dear Ivanushka!' cried the Princess. 'I shall give you back your sword!'

And sobbing hysterically, she ran into the next room and returned with the magic sword. Ivan grasped it by the hilt, and running quickly out of the room, he found the foreign prince in the courtyard, and with one blow struck off his evil head.

'This is my sword, all right,' muttered Ivan contentedly, and strode back to the Princess's room. There he took two red berries from his pocket and gave them to her to eat. The Princess ate one berry, and one of the horns fell to the ground; she ate a second berry, and the second horn fell off, and she began to laugh and cry for joy.

'Thank you, thank you, dear Ivanushka!' she cried. 'You have delivered me from misfortune a second time, and I shall never forget your kindness. Drive the foreign prince away, forgive me all my sins, and suffer me to remain your faithful wife for ever.'

'The foreign prince is already where you ought to be,' said Ivan. 'You are unfit to be the wife of any man, just as your father and mother are quite unworthy to be King and Queen. You are all without honour, and shall be banished for ever from this land!'

And so Ivan the Soldier drove the perfidious King, Queen, and Princess from the Kingdom and ruled in their stead, and with the Samosek Sword to defend it, the Kingdom prospered ever after.

The Firebird

THE magnificent gardens that surrounded the palace of Tsar Vyslav Andronovich were so full of rare and wondrous flowers and trees that there were no better to be found in the whole wide world, and in these gardens grew a splendid tree, specially dear to the King's heart, which bore golden apples. One morning, however, the King saw to his dismay that his favourite tree had suffered much damage during the night, for many apples were missing from its branches, while others lay scattered about the ground. The Tsar was sorely affected by this wanton destruction, and hastily summoning his three sons Dimitri-Tsarevich, Vasili-Tsarevich, and Ivan-Tsarevich into his presence, he said:

'My beloved children, you have seen the sorry state of the golden apple-tree which I treasure so much, and it is obvious that something or somebody has intruded into my grounds during the night to wreak this havoc. This must be stopped, and to this end I hereby solemnly promise that whosoever of you succeed in capturing the culprit shall receive half of my Kingdom in my lifetime and the whole of it after my death.'

'Your Majesty,' replied his sons, 'we shall each of us endeavour to catch the culprit alive.'

On the first night it was the eldest prince, Dimitri-

189

Tsarevich, who went into the garden to keep watch, but no sooner had he arrived beneath the golden apple-tree than he settled himself comfortably down against the tree-trunk and went to sleep. In the morning, when he awoke, the tree was in an even worse state than before, and the Prince had to confess to his angry father that he had neither seen nor heard anything all night. It was the same story the second night when the second son, Vasili-Tsarevich, went to keep watch; the intruder came while he was asleep and he saw nothing, and the next morning he thought fit to try to persuade his angry father against the evidence of his eyes that the thief had not come at all. The third night it was the turn of the youngest prince, Ivan-Tsarevich, to keep watch in the garden, and hiding beneath the golden apple-tree, he did not go to sleep but watched attentively. One hour passed and then another, when suddenly, at the end of the third hour, the whole garden was filled with the clear light of day, and in front of the eyes of the watching prince a large and beautiful bird, its feathers shining like gold and its eyes like crystal, flew swiftly into the garden, alighted gracefully on the branches of the golden apple-tree, and began to tear away at the fruit with its amber beak. It was the magic Firebird! Ivan, however, was quick to recover from his astonishment, and so skilfully did he creep up upon his quarry that the bird did not hear him until it was almost too late. Ivan's hand closed round the Firebird's tail, but the frightened bird started up in panic, wrenched itself from the Prince's grip, and flew swiftly away. The bird, however, had not escaped entirely unscathed, for in Ivan's hand—for he had held on with all his might—remained one sole feather from its tail.

When Tsar Vyslav Andronovich summoned his son Ivan-Tsarevich the following morning and heard his account of the events of the preceding night, he was so impressed by the beauty of the single tail feather which still shone like the very sun in heaven, that he stored it away in his jewel cabinet as a great treasure and eagerly looked forward to the capture of the Firebird.

Alas! from that day forth the Firebird came no more to Tsar Vyslav's garden, and though the golden apple-tree recovered and flourished again in all its former splendour, the King was sad. He waited in vain for several weeks, and then, summoning his two elder sons into his presence, he promised to give his whole kingdom forthwith and without delay to whichever of his sons should bring him back the magic Firebird. Now Dimitri-Tsarevich and Vasili-Tsarevich had borne great ill-will against their younger brother Ivan-Tsarevich ever since he had succeeded in winning his father's esteem by snatching the golden feather from the magic bird's tail, and they were only too pleased to set out together with their father's blessing in search of the Firebird. Ivan-Tsarevich was sorely disappointed that his father had neglected to summon him, and when his patience could bear it no longer, he went to the Tsar and asked for his blessing, for he too would depart in search of the bird.

'My beloved son,' said the Tsar, 'you are still young and unaccustomed to long and difficult journeys. If you go, I may never see any of my sons alive again, for I am old and journey fast towards God. If the Lord take my life, who shall rule the kingdom in my stead? If unrest and rebellion break loose among the people, who shall put it down? If a foreign foe attack our land, who shall lead the army against them?'

But though Tsar Vyslav Andronovich sought hard to restrain his youngest son, all was of no avail; he was finally persuaded to give his blessing, and Ivan-Tsarevich mounted his horse and rode away he knew not whither.

He rode near and far and high and low—words are swift but deeds are slow—until he came to a green meadow. And there in the green meadow stood a wooden post on which these words were written:

'He who goeth straight on shall hunger and thirst; he who goeth to the right shall prosper, but his steed shall perish; he who goeth to the left shall perish, but his steed shall prosper.'

Ivan-Tsarevich spurred his horse along the road to the right, and when he had journeyed one day and then another, a huge grey wolf suddenly darted towards him on the third.

'Ho there, Ivan-Tsarevich!' cried the wolf. 'You have read what is written on the post!'

And with that, the wolf leapt ferociously at the throat of Ivan's horse, gobbled it up, and sped swiftly away.

Ivan wept bitterly over the death of his horse, for he was very fond of it, but he plucked up his courage and continued his way on foot. When he had walked a whole day thus and sank wearily down to rest a while, the grey wolf suddenly came bounding towards him.

'Forgive me for devouring your steed, Ivan-Tsarevich,' he said, 'but such is the nature of a wolf. But I grieve to see you footsore and weary from much walking, and if you sit on my back, I will gladly carry you wherever you want to go.'

Ivan-Tsarevich settled himself gratefully upon the broad back of the huge grey wolf, told him the reason for his journey, and the wolf sped swiftly over Damp Mother Earth. Within the space of one night they arrived at the foot of a tall stone wall and the wolf bade the Prince climb over the wall into the garden that lay beyond.

'For there,' said the wolf, 'you shall find the Firebird hanging in its golden cage. Take the Firebird out, but whatever you do, do not touch the cage itself, or you will surely be captured.'

Ivan did as he was bidden, and climbing over the wall, he beheld the wonderful Firebird in its golden cage. For a moment he stood motionless, enraptured by the bird's transcendent beauty, but he set his mind on his task, walked up to the cage, removed the Firebird, and walked back to the wall. Then he stopped.

'I have the Firebird,' he thought, 'but where shall I put it?'

So boldly ignoring the wolf's friendly warning, Ivan went back and took the golden cage from the branch on which it hung. No sooner had he done this, however, than a loud commotion broke out on all sides, for hidden strings led from the cage to the many bells that now sounded the alarm. The sentinels awoke, ran to the garden, and falling upon Ivan-Tsarevich, dragged him unceremoniously before their master, the mighty King Dolmat.

'What manner of man are you,' cried the angry King, 'that you climb like a thief into my grounds and steal my property? What country do you come from and what is your name?'

The Prince hung his head in shame.

'I am the son of Tsar Vyslav Andronovich and am called Ivan-Tsarevich,' he said. 'My father sent me in search of your Firebird, for every night it would fly into our garden and wreak havoc on our golden apple-tree.'

When he heard this, the King's tone softened.

'Ivan-Tsarevich,' he said, 'why did you come to me like a thief in the night? If you had come to me and stated your honoured father's wishes, I should willingly have given you the magic Firebird. Now, however, I must spread the news of your dishonour throughout all the lands of the earth, and what shall men say of Ivan-Tsarevich?'

The King paused, and Ivan hung his head more deeply in shame.

'However,' continued King Dolmat, 'if you are willing to undertake a great errand on my behalf, I shall say nothing of the whole matter and shall give you the Firebird as a reward for your services.'

'What is the errand, O King?' asked Ivan-Tsarevich, willing to face death rather than dishonour.

'You must ride to the Kingdom of King Afron in Thrice-Ninth Land, and bring back his horse with the golden mane. Do this, and you shall be pardoned and handsomely re-warded! Fail, and I shall tell the world of your dishonour-able theft!'

Ivan-Tsarevich had no choice. He sadly took his leave of King Dolmat, returned to his friend the grey wolf and told him all that had happened.

'Did I not tell you it was dangerous to touch the golden cage?' said the grey wolf.

'Forgive me, grey wolf,' said Ivan-Tsarevich. 'I was a fool to ignore your warning.'

'Never mind, then,' replied the wolf. 'Mount on my back, and I will carry you wherever you want to go.'

The next night the wolf and Ivan-Tsarevich passed into the land of Thrice-Nine and came to a halt before the white stone stables of King Afron.

'Go into the stables,' said the wolf, 'and lead out the horse with the golden mane. But whatever you do, do not touch the golden bridle hanging on the wall, or you will surely regret it.'

Ivan did as he was bidden and was leading the wonderful white horse with the golden mane from the stables, when his eye caught the golden bridle hanging on the wall.

'How can I ride the horse without a bridle?' he thought, and reaching out his hand he made to take it down from the wall. But the bridle also was joined by hidden strings to secret bells, and in no time he was seized by the royal grooms and dragged before the mighty King Afron of Thrice-Ninth Land.

'Who are you,' cried the angry King, 'that you come creeping up like a horse-thief into my stables? What country do you come from and who is your father?'

'I come from the land of Tsar Vyslav Andronovich,' replied the Prince, hanging his head in shame, 'and my name is Ivan-Tsarevich.'

'Alas, young Ivan-Tsarevich!' replied King Afron. 'This is no deed of a noble knight. If you had come to me openly, I should have given you the horse as a gift, but now I am compelled to tell the world of your dishonourable actions. If, however, you are willing to perform a little service for me, I shall say nothing of this affair and shall give you the horse with the golden mane as a reward. Ride beyond the Land of Thrice-Nine and bring me back Fair Helen, the daughter of the King, for long have I been enamoured of this maid. Do this for me, and you shall be pardoned and richly rewarded! Fail, and I shall tell the world you are a thief!'

There was nothing for Ivan-Tsarevich to do but to agree to the King's request, and taking his leave of the King, he returned to the grey wolf and told him all that had happened.

'Why did you not pay heed to my words?' cried the wolf. 'Did I not tell you not to touch the golden bridle?'

'I am sorry, grey wolf,' replied Ivan. 'Please forgive me.'

'Very well,' said the wolf. 'Mount on my back, and I will carry you whither you want to go.'

Ivan-Tsarevich mounted the grey wolf's back a third time and the animal sped like an arrow through the night until they arrived in the land where Fair Helen dwelt with her father the King. This time the grey wolf bade Ivan wait for him in a green meadow beneath an oak-tree while he himself went in search of the Princess. Ivan did as he was bidden and the wolf ran off and halted before a wooden fence. Now it was at this hour, when the red sun was setting in the West and the air grew cool after the heat of the day, that the beautiful Princess was wont to walk abroad in the gardens with her ladies-in-waiting, and no sooner had she drawn near to the fence than the grey wolf leapt over, seized Fair Helen, and made off with her, back to where Ivan-Tsarevich was waiting in the meadow. The wolf bade him quickly mount his back beside Fair Helen, for the Princess's suite had given the alarm and the pursuers could already be heard in the distance. Ivan did as he was told, and together they rode back towards the palace of King Afron, while the sounds of the horsemen who galloped in pursuit grew fainter and fainter and were soon left far behind.

As they rode along on the grey wolf's back Ivan-Tsarevich and Fair Helen fell deeply in love one with the other, and when eventually they arrived in the land of King Afron, the Prince began to weep.

'Why do you weep, Ivan?' asked the wolf.

'Alas, dear friend!' sobbed Ivan. 'How should I not weep? I am in love with the fair Princess and she with me, and now the time has come for us to part! For if I fail to hand her over to King Afron in exchange for the horse with

the golden mane, he will brand me as a thief throughout all
the countries of the world.'

'I have served you well for the price of your horse, Ivan-
Tsarevich,' said the grey wolf, 'and I shall not fail to do you
yet another service. I shall change myself into a maiden, the
living image of Fair Helen. You shall lead me to King
Afron and take the horse with the golden mane in exchange.
The king will honour me as a real princess, and in the meantime you must mount the horse beside Fair Helen and ride
away as fast as you can. I shall then request King Afron to
suffer me to walk a while in the garden with my ladies-in-
waiting, and as soon as I am outside the palace, you have
only to think of me and I shall straightway reassume my
true form and return to you.'

Hereupon, the grey wolf struck the damp earth and
changed forthwith into a beautiful princess so like Fair
Helen that none could have told them apart. Telling Fair
Helen to await his return, Ivan-Tsarevich took the false
princess by the hand and led her to the palace of King Afron,
who, much taken by her radiant beauty, gladly and un-
hesitatingly gave the Prince the horse with the golden mane—

and the golden bridle—in exchange for her. Then Ivan-Tsarevich mounted the horse, and riding back to where Fair Helen awaited his return, he lifted her up in the saddle beside him and galloped swiftly away.

For one day, and then another, and yet a third, the grey wolf lived in King Afron's palace in the guise of the fair Princess. He was treated with great honour and deference, and when on the fourth day he asked the King if he might walk in the palace gardens with the ladies-in-waiting, the request was readily granted.

'Fair Helen' said the King, 'is there anything in the world I would not do for you?'

And King Afron commanded all the ladies and governesses and maidservants to accompany the Princess and watch that she came to no harm.

In the meantime, Ivan-Tsarevich was riding leisurely along, chatting and laughing with Fair Helen and quite unmindful of his friend the grey wolf. Luckily, however, Fair Helen reminded him of the wolf's instructions, and no sooner did the Prince's mind turn to the wolf, than far away behind them, the false princess suddenly quit her human form and turned into a huge grey wolf. The ladies-in-waiting, the governesses, and the maids screamed in terror and fainted right away, and when they eventually came to their senses, the wolf had fled. All that remained was the royal cloak given her by King Afron, and they all assumed that their mistress had been gobbled up by a big, grey wolf. What King Afron had to say about this is not told in the story, but in any case, before he could express any opinions on the subject, the grey wolf was standing once again before Ivan-Tsarevich.

'Sit on my back, Ivan!' said the wolf, 'and let Fair Helen ride on the horse with the golden mane.'

And so they rode on thus to the lands of King Dolmat, but the nearer they drew, the more worried grew the young Prince's brow.

'Listen, Grey Wolf,' he said at last. 'I am exceedingly loth to part with this wonderful horse to whom I have taken a

great fancy. If you could change into a horse as easily as you
can into a princess . . .'

Here he broke off, for the grey wolf had struck the damp
earth and in his stead stood a fine horse, the very image of
the horse with the golden mane.

Bidding Fair Helen again await his return, Ivan sprang
on to the back of the magic horse and galloped away to the
palace of King Dolmat, who, delighted beyond all measure
at the sight of what he supposed to be the real horse with
the golden mane, gladly gave Ivan-Tsarevich the Firebird
in exchange. As soon as Ivan-Tsarevich had regained Fair
Helen, however, and thought of the wolf, there before the
horrified King Dolmat's eyes the much-coveted steed suddenly turned into a huge grey wolf.

In a flash the wolf was back at Ivan's side. The Prince
again mounted the wolf's back and thus, accompanied by
Fair Helen, the horse with the golden mane, and the magic
Firebird, they journeyed back towards the lands of his
father, Tsar Vyslav Andronovich. When they came to the
spot where lay the white bones of Ivan's old steed, however,
the wolf halted.

'Well, Prince,' he said, 'we are now back at the place
where we first met. Here it was that I ate up your horse,
but for that misdeed I have, I think, made amends. I have
served you well, and you now have everything you could
possibly wish for. Mount the horse with the golden mane and
ride back to your father with Fair Helen, for my task is done.'

And saying this, the grey wolf turned about and sped
swiftly away. Ivan was dumbfounded and began to weep
bitterly for the loss of his trusty friend, but seeing there was
nothing to be done, for the wolf had already disappeared
from sight in the open plain, Ivan mounted the horse with
the golden mane beside the fair Princess and rode off in the
direction of his father's palace. When they were within
thirty versts of their destination, however, the sun began to
beat down mercilessly upon their heads, and dismounting
from the horse Fair Helen and Ivan-Tsarevich lay down in
the shade of a tall tree and fell asleep.

Now just at this time the Prince's two elder brothers, Dimitri-Tsarevich and Vasili-Tsarevich, were returning empty-handed from their fruitless quest, and passing along the road that led to their father's palace, they caught sight of the group resting beneath the tree.

'Why, it is Ivan-Tsarevich, our younger brother!' whispered Vasili. 'And not only has he got the magic Firebird, but he has got a beautiful princess and a horse with a golden mane as well!'

'This is too much!' muttered Dimitri. 'A younger brother does not deserve such luck. A quick thrust of the sword and the treasures are ours!'

And so, creeping up behind the tree, the wicked Dimitri drew his sword from his scabbard and slew his younger brother on the spot. Then the two murderous princes hacked the body of their unfortunate brother into little pieces, and when this was done, they awakened the sleeping Princess.

'Wake up, fair maiden!' they said. 'Tell us where you come from, whose daughter you are, and what your name is!'

When the Princess saw the mutilated body of her beloved Ivan-Tsarevich, she trembled violently and burst into tears.

'I am called Fair Helen and am the daughter of a King,' she sobbed. 'Ivan-Tsarevich, whom you have slain, brought me hither. But what manner of men are you? Had you been noble knights you would not have slain a sleeping man, for he who sleeps is as one dead!'

Dimitri-Tsarevich only laughed cruelly.

'Listen to me, Fair Helen!' he said, pointing his sharp sword at the Princess's breast. 'You are a prisoner in our hands, and we shall take you to our father, Tsar Vyslav Andronovich. To him you shall declare that we, Dimitri and Vasili, brought you hither, and the Firebird, and the horse with the golden mane. If you refuse, we shall kill you here and now!'

Fair Helen was terrified at the sight of the bloody sword and promised to do as they commanded. The two brothers threw lots for the booty, Vasili winning the Princess and Dimitri the horse. Then, mounting their horses, they rode back to their father's palace to tell their false tales of heroic deeds.

For ninety days after the murder the body of Ivan-Tsarevich lay rotting in the place where he was slain, but on the ninety-first day the grey wolf chanced to pass by the tall tree, and when he recognized his friend Ivan, he was stricken with grief and mourned deeply. Gladly would he have brought his young friend back to life, but this was a secret that not even he possessed. Just at that moment, however, he heard a flapping of wings, and hiding behind the tree, he saw an old crow and two young crows fly down to feed off the flesh of Ivan's corpse. No sooner had their sharp beaks touched his body, however, than the grey wolf darted out from his hiding-place, and seizing one of the two young crows in his sharp teeth, made to rip him in two.

'Grey Wolf, Grey Wolf!' cried the old crow. 'Do not harm my child! He has done you no wrong!'

'Listen to me, Voron Voronich!'* replied the wolf. 'Fly to the Thrice-Nine Land and bring me back the Waters of Life and Death. Do this for me, and I shall release your child safe and sound. Refuse, and I shall tear him to shreds!'

'I shall do you this service, Grey Wolf!' cried the crow, 'only do no harm to my child!'

The old crow flew swiftly away, and at the end of three days he returned carrying two small flasks in his beak. The wolf took the two bottles, and before the old crow could protest, he had torn the young one apart. Then the wolf sprinkled a drop of the Water of Death over the body and immediately the two halves grew together again; then he sprinkled a drop of the Water of Life over the body, and straightway the young crow came to life again and flew off with the two others. Then the grey wolf performed the same rites over the body of the Prince, and in less time than it takes to tell, Ivan-Tsarevich was sitting up rubbing his eyes.

'What a long sleep I have had!' he exclaimed.

* Crow, son of Crow.

'You would have slept for ever had it not been for me!' replied the grey wolf. 'Your two brothers slew you in your sleep, robbed you of the Princess, the Firebird, and the horse with the golden mane, and have told your father that it was they who brought back all these wonderful things. But jump on my back and let us make haste to your father's palace, for today Vasili-Tsarevich is to take Fair Helen for his bride.'

Ivan needed no second bidding. He leapt on to the wolf's broad back, and they sped like the wind to Tsar Vyslav's palace. Here Ivan dismounted from the wolf's back, and running into the hall, he found the marriage feast was already prepared. Fair Helen was sitting in her bridal robes, but as soon as she spied her true lover, she gave a cry of joy, rose from the table, and rushed into Ivan's arms.

'This is my true betrothed!' she cried. 'And that scoundrel who sits there and claims to be my bridegroom was his would-be murderer!'

And thereupon Fair Helen told the Tsar of all that had really happened. The Tsar was greatly shocked to hear of the callous cruelty of his two elder sons, and angrily commanded them to be cast into chains and thrust into the deep dungeons below. Ivan-Tsarevich married the Princess and inherited half his father's kingdom, and all, Ivan-Tsarevich, Fair Helen, Tsar Vyslav Andronovich, the Firebird, and the horse with the golden mane—all, in fact, except the murderous impostors Vasili and Dimitri—lived happily together in peace and prosperity.

The Hired Man

I N a small village near Moscow there once lived an old
muzhik who had three sons. When they grew up, the
eldest went into the town to hire himself as a labourer
to a merchant who was as mean as he was rich.

'When the cock crows, my lad,' he said to the eldest son,
'you must get up and start work at once.'

For three days the eldest son suffered this harsh regime,
but on the fourth day he failed to hear the cock crow and was
straightway dismissed without a copeck.

Then the muzhik's second son went into the town and
hired himself as a labourer to the same merchant, but when
he had worked for six days, he failed to hear the cock crow
on the seventh, and he too was dismissed.

When the youngest son told his father that he too would
like to try his luck with the rich merchant, his brothers,
whose lack of success had not increased their modesty, burst
out laughing.

'A lot of good it will be to send him, that's sure!' they
cried. 'But he might as well go. At least there'll be a little
more room on the stove for a day or so perhaps, with any
luck!'

So the youngest son went and hired himself out to the
merchant.

'When the cock crows, my lad,' said the merchant, 'you must get up and start work at once, and stay at it till nightfall!'

'All right,' said the lad.

'And what is the price of your hire?' asked the merchant.

'Oh, I don't need much,' replied the young man. 'Let us say that if I work a whole year for you I shall have the right to give you a punch and your wife a pinch.'

'Oho!' chuckled the merchant to himself. 'Here's a green one, all right!'

And so the bargain was struck.

The same evening, after the merchant had retired to bed, the hired man went into the yard, grabbed hold of the cock and tied its head under its wing. Then he went to sleep.

When dawn broke, no cock crowed, and it was not until the sun was high in the heavens that the labourer awoke, rubbed his eyes, and asked for his breakfast.

'Well, merchant,' he said. 'I cannot wait all day for the cock to crow. It's time to get to work.'

The merchant could not think what had happened to the bird, and he decided to go into town with his labourer to buy another one. As they walked along, they met four men with a huge ox which was so strong and fiery that the ropes were strained to breaking point and it was all the men could do to keep up with it.

'Where are you going, brothers?' asked the labourer.

'We are taking the ox to be slaughtered,' they replied. 'But it is giving us a great deal of trouble.'

'Why not slaughter it here and now?'

'We have no hammer heavy enough,' said the men, but going up to the ox, the hired man raised his fist, and gave the animal such a blow on its forehead that it fell down stone dead.

'That has done the trick,' said the ox-herds. 'But now we shall have to drag it into town to be skinned. We have no pincers with us.'

But the labourer stooped down, and pinching the ox in

the rump between forefinger and thumb, he pulled off the whole hide as though it were the skin of a ripe fig.

When the merchant saw the quality of the young man's punches and pinches, he realized what a dreadful bargain he had made, and forgetting all about the new cock, he returned home and related the whole affair to his wife.

'I know just what to do!' said the latter, who was as sly and mean as her husband. 'Tonight we shall tell him that one of the cows has failed to return with the herd. Then we'll send him into the dark forest to look for it, and the wild beasts will eat him up.'

After supper, the merchant went to the hired man's hut. 'Why didn't you count the cows when they returned?' he said angrily. 'One is missing from the herd.'

'One missing?' repeated the young man. 'But I did count them.'

'Then you counted them wrong, fool!' snapped the merchant. 'Now go and look for the missing one!'

The young man took his wooden club and walked out into the dark forest. For a whole hour he searched in vain, but at last he saw a large black shape lurking in a cave. Now this was a bear, but in the darkness he took it to be the missing cow.

'Aha, you wretch!' he cried. 'You certainly hid yourself well. But now, come here!'

And he began to lay about the bear with his wooden club, and though the bear scratched frantically at the walls of its den in a vain attempt to escape, the young man seized him by the scruff of the neck, dragged him back to the farm, and kicked him into the cowshed.

The next morning he went to the merchant and told him he had found the missing cow.

'What cow could the fool have found?' wondered the merchant, and when he opened the door of the cowshed, all the cows lay dead on the ground in a pool of blood and bones, and the bear snored contentedly away in the corner.

'Fool!' screamed the merchant. 'You have put a bear in among the cows, and he has killed the lot!'

'So it appears,' said the young man. 'But he'll suffer for this!'

And rolling up his sleeve, he punched the bear between the eyes, and its life went out like a light.

'The state of affairs gets worse and worse!' thought the merchant. 'Wild beasts can do nothing with him, but let us see what a few devils can do!'

And aloud to the hired man he said:

'Go to Devils' Mill and collect all the money they owe me. The mill is on my land, and they owe me years of rent.'

The hired man took a horse and cart and drove up to the Devils' Mill, and sitting on the wall, he began to unroll a large coil of rope.

A devil's head popped out of the water and eyed him curiously.

'What are you doing, brother?' he asked.

'I am unwinding this rope.'

'What do you want the rope for?'

'I am going to rope all you devils together and drag you out of the water and shrivel you up in the sun.'

'Why do you mean us harm, brother? We have done you no wrong.'

'Why haven't you paid my master what you owe him?' said the young man grimly.

'Wait a minute, brother, don't do anything rash,' pleaded the devil. 'I'll go and ask the chief!'

The devil's head popped back beneath the surface of the water, and taking a spade from his cart, the labourer dug a large pit which he covered with dry branches. Then he took off his cap, cut a hole in the middle and placed it on top of the branches over the pit.

A few minutes later, the devil's head reappeared.

'The chief asks how you intend to drag us out on the end of a rope when our pit is bottomless?' said the devil, full of smiles.

'Why, with this endless rope of mine, of course.'

'That is impossible! I'll soon find the end!'

But the labourer had already plaited the two ends of the

rope together, and however much the devil drew it through his fingers, it never ended. The devil gave up and scratched his head.

'Do we owe you much?' he asked.

'Oh, just fill my cap with silver,' replied the young man, 'and I'll leave you in peace.'

The devil disappeared into the mill-pool again and related everything to his superior. It was hard to have to part with the silver, but there was nothing for it, and the young devil was authorized to fill the labourer's cap. But, of course, the coins fell straight through the hole in the cap into the deep pit hidden beneath, and it was a very large amount of silver indeed that the labourer finally loaded on to his cart and took back to the merchant.

The greedy merchant, however, viewed his success with mixed feelings.

'The silver is good,' he thought, 'but it does not please me to know that the devils themselves cannot get the better of this man. Moreover, the time is fast approaching when he will demand his due, and then I shall be pole-axed and my wife flayed alive!'

These thoughts weighed so heavily upon his mind that, in desperation, he decided to take flight. His wife baked two large sackfuls of pastries so that they should not go hungry on the way, and they waited impatiently for night to fall. Their labourer, however, had noticed these preparations, and when the merchant and his wife were not looking, he crept into the kitchen, emptied the pastries out of the sacks, and putting a millstone in one, concealed himself in the other.

When night came, the merchant and his wife took each a sack, threw it over their shoulders, and made off as fast as their legs would carry them. They had not gone far, however, when they heard the muffled voice of their labourer behind them.

'Hey, master! Mistress! Take me with you!'

The frightened couple nearly fell dead in their tracks!

'He's found out, curse him!' said the merchant, and not

suspecting for a moment that the voice came from one of the sacks they carried on their backs, they ran on faster and faster.

'Uff!' panted the merchant, when they came to a lake. 'Let us take a rest. These bags are very heavy.'

And so saying, he threw his sack roughly to the ground.

'Ow!' cried the labourer. 'Gently, master, or you'll break my ribs!'

He stepped out of the sack.

'Heaven help us!' cried the merchant. 'So you have been here all the time?'

'Yes, master. I did not want you to leave me behind.'

Well, once again there was nothing to be done, and all three lay down to pass the night at the side of the lake, but when the merchant heard his labourer breathing heavily, he whispered to his wife:

'Let us wait till the fellow is soundly asleep, and then we'll throw him in the lake. But go to sleep now, and we'll wake up later.'

But the hired man had heard their whispered conversation, and when they themselves were fast asleep, he got up, covered the wife with his old cap and sheepskin coat, dressed himself in her fur coat, and then woke the merchant.

'Come on, husband!' he whispered. 'Let us throw him in now!'

The merchant rose quickly, seized his wife by the feet while the hired man took her head, and together they threw her into the lake, where, with a loud splash and a gurgle, she sank to the bottom.

'Well, that's that!' said the labourer, standing up straight and speaking aloud in his normal voice. 'But are you sure you wanted to drown her?'

The merchant's eyes bulged out of his head as the truth slowly dawned upon him, and he returned sadly back to his farm with his hired labourer. When the year of service was up, the young man gave his master the promised punch on the head, and that was the end of him. The young man then took over the property and lived happily ever after.

The Fortunate Maiden

ONCE upon a time there dwelt in the town of Kiev a rich merchant and his son, and when the lad was old enough to take a bride, his father called him and said:

'Go, my son, to the rich merchant's daughter and ask for her hand. Have no fear: she will not refuse you, for I have arranged everything with her father.'

The young man, however, shifted and shuffled about and looked miserable.

'Father,' he replied, 'I should much rather go out into the world and look around for myself, for I might well find a bride far richer and fairer than she.'

The father did not insist, and receiving permission to seek his own fortune in the world for the space of one year, the youth set out. On and on he walked until it began to get dark, and when he looked about him, he found that he had lost himself in a dark wood, and there on his right stood a tiny hut supported on fowls' feet. Approaching the hut, the lad knocked on the door and went in. Inside sat a very old man with a beard as white as milk. It was Fate himself.

'Good evening, batiushka,' said the traveller.

'Good evening,' replied Fate. 'What brings you to my lowly hut?'

'I lost my way in the forest, and seeing your hut, I took the liberty of entering,' replied the young man. And looking about him, he saw that there were three stools in the hut— one gold, one silver, and one brass.

'What manner of stools are they?' he asked.

'They are stools for me to sit on, of course,' replied Fate. 'But the man who is born whilst I am sitting on the gold stool will be exceedingly lucky in life; he who is born whilst I am sitting on the silver stool will be moderately lucky; and he who is born whilst I am sitting on the brass stool will be unlucky.'

'What stool were you sitting on when I was born?' asked the young man.

'On the brass one.'

'And when the rich merchant's daughter was born?'

'On the brass one.'

When Fate saw the gloom that spread over the young man's face, he took pity and added:

'But whilst I was sitting on the silver stool, the girl who now carries water for the Jews was born. Marry her and you will both be very lucky.'

The lad thanked the old man for his advice and returned to his father, who asked him whether he had found a fitting bride.

'I have, Father!' replied the lad. 'Give me your blessing, for I have decided to marry the girl who now carries water for the Jews.'

'What?' cried his father, unable to believe his ears. 'Have you gone mad? If you marry the handmaid of the Jews, it is not my blessing I shall give you, but my curse!'

But the young man insisted, and his father angrily ordered him out of the house and told him never to return again. The young man left his father, went to the Jews' handmaid, gave her some clothes, and married her. Everyone laughed at them except the father, who wept bitterly to see the folly of his only son.

The day after the wedding, the young man gave his new wife fifteen roubles and told her to go to the village to buy enough wares to set up a shop and trade. The wife went to the market-place and returned with a cart-load of coal, and when her husband saw that this was all she had bought for the money he had given her, he almost wept.

'What a fool I was to ignore my father's advice and listen to that old man,' he thought bitterly. 'We shall never be lucky.'

In the evening, however, when he went to fetch some coal for the stove, he found that all the coal had gone, and there in its place lay a pile of gold coins.

And so the young man was reconciled with his father, and all three spent their lives in comfort and luxury.

> I was their guest, you know,
> And mead and wine drank I.
> It flowed right richly through my beard
> And yet my mouth stayed dry.

The Wife who liked Fairy-tales

A MAN once married a wife who had such a passion for fairy-tales that she resolutely refused to admit anybody into the house who was unable to tell her any. Her husband soon exhausted his meagre store, and thereafter had to sit and listen to all the old babushki in the village as they gathered around his wife, drinking his tea and eating his pirozhki and telling her all manner of nonsense about this prince and that dragon for days on end. As you can well imagine, the poor fellow was soon at his wits' end. No friend of his could ever set foot in the house, for as soon as his wife knew that he was not much of a one at telling stories, she would plant herself in the doorway and refuse to let him in.

Well, one winter's evening in December when the snow lay thick on the ground and icicles hung from the frosted window-panes, an old man, shivering from the cold, knocked at the door of their hut and begged for shelter for the night.

'What are you like at fairy-tales, old man?' asked the husband with a worried look. 'I myself would gladly give you shelter for the night, but the fact is, my wife will let no one into the house who cannot tell her some story or other.'

The old man was so tormented by the cold that he would

have undertaken almost anything to be allowed near the warmth of a good stove, and he solemnly declared that there was none in the whole of Russia who knew more fairy-tales than he—a claim which, to tell the truth, was itself a fairy-tale.

'Well, if you want to stay the whole night here,' said the husband, 'you will have to tell your stories the whole night through!'

'All right,' said the old man. 'But I must point out that I am a real artist and that we artists have most sensitive natures. I must therefore insist that there shall be no interruptions while I am telling my tales!'

And so the old man was welcomed into the house and began to warm himself in front of the warm stove.

'Now, wife!' said the husband, 'our guest will tell you his stories only on condition that you promise not to interrupt him while he is speaking. If you do, he will not continue.'

'I interrupt?' cried the wife, delighted at the thought of eight glorious hours of fairy-tales. 'If I interrupt him, I swear never to listen to another fairy-tale again! It is you he ought to be worried about, not me!'

Supper was prepared and eaten, and they settled themselves comfortably on top of the stove to listen to the old man.

The old man coughed and cleared his throat and shut his eyes as though in a trance, while the wife cupped her hand to her ear, anxious not to miss a single word. The old man began to tell his tale:

'An owl flew into a garden, alighted on a well, and drank some water; an owl flew into a garden, alighted on a well, and drank some water; an owl flew into a garden, alighted on a well, and drank some water; an owl flew into a garden, alighted on a well, and drank some water . . .'

The old man's voice droned on and on, and though his voice was strong and full of feeling, it was always the one and the same phrase over and over again:

'An owl flew into a garden, alighted on a well, and drank some water . . .'

At last the wife realized with some impatience that there were more repetitions in this particular fairy-tale than in any she had ever heard before, but she listened and listened and listened until her ears could bear it no longer.

'What sort of a tale is this?' she cried. '*An owl flew into a garden, alighted on a well, and drank some water,* indeed! It's always one and the same thing!'

'Why have you interrupted me?' cried the old man angrily. 'The prologue is a most essential part of a fairy-tale and is designed to put one in the proper frame of mind, and that is how this particular story starts off! The rest would have followed later, but since you have broken your promise not to interrupt me, I decline to continue!'

'Now we shall never know what happened to that owl!' cried the husband angrily, leaping down from the stove and beating his wife. 'Did you not promise to listen attentively to our guest and swear not to interrupt him? Since you have shown no interest in this fairy-tale, I vow that you shall never hear another one in your whole life—at least not in this house!'

And beating his wife so mercilessly that she never wished to hear another fairy-tale for as long as she lived, the husband secretly rewarded the old man for his presence of mind.

Glossary

arshin	28 inches
ataman	Cossack or robber chieftain
babushka	grandmother
batiushka	father, old man
bogatyr	hero, knight
Boyan	an ancient bard
boyar	nobleman
druzhina	bodyguard, retinue
Gospodi!	Lord!
gusly (pr. goosly)	psaltery, zither
izba	peasant's hut
leshi	woodland sprite
Likho	Evil; personified as a witch
mirza	Tatar prince
muzhik	peasant
pirozhki	pastries
polianitsa	female warrior
pood	36 pounds
Tsargrad	Constantinople
ulan	Tatar lancer
-ushka	diminutive suffix
vedro	bucketful ($2\frac{3}{4}$ gallons)
verst	about $\frac{2}{3}$ of a mile
yaga	witch (as in Baba Yaga)

Acknowledgements

The author would like to acknowledge the following as the sources of the stories contained in this book:

ONEŽSKIE BYLINY, collected by A. F. Gilferding. 4th edition. Moscow–Leningrad, 1949.

BYLINY SEVERA, ed. A. M. Astakhova. Moscow–Leningrad, 1938.

BYLINY, ed. M. Speranski. Moscow, 1916.

NARODNYE RUSSKIE SKAZKI I LEGENDY, collected by A. N. Afanas'ev. Berlin, 1922.